Birth of a Breed

Published in cooperation with the
AMERICAN POLLED HEREFORD ASSOCIATION
by THE LOWELL PRESS, Kansas City, Missouri

Birth of a Breed

The History of Polled Herefords— America's First Beef Breed

by ORVILLE K. SWEET

This is the RANCH EDITION, which is
limited to 100 custom bound copies. It
has been numbered and signed by
Orville K. Sweet.

Book Number *1* of 100

First Edition
©Copyright 1975 by American Polled Hereford Association
All Rights Reserved

Printed in the United States of America
Library of Congress Catalog Card No. 75-18757
ISBN 0-913504-25-4

Dedication

I dedicate this book to the wives of the members of the Polled Hereford Hall of Fame, who deserve special recognition and equal honor with their husbands.

I also dedicate this effort to my mother, Mary, and to my wife, Lew, who not only has been an inspiration during my years with the breed but has shown a special kind of tolerance for my coming and going, restless nights, and the early-hour disturbances, while I collected these thoughts.

Foreword

The story of beef cattle in America and the people who lived this epic is unparalleled in color, romance, and tradition. The colorful cowboy, that hero of early America, has stirred the imagination of old and young around the world. No other professional, not even the legendary soldier, the dedicated doctor, or the respected lawyer, has charged the emotions or appealed so strongly to the hero worship that lies in us all. That may have been Warren Gammon's motivation. For it was in the mind of that Iowa country lawyer that Polled Herefords began simply as a good idea and finally emerged as a dream that became a reality.

The cowman was respected, envied, and glorified for what he did, not for what he had. The cowboy, whose tools were a rope, a running iron, and a tough, individualistic outlook on life, has evolved into a modern-day technician. He is now armed with a degree in animal science and the ability to measure and manipulate genetic material he cannot see, and he is guided by an enthusiastic objective to improve the economic value of the cow. The early-day cattleman simply ran a herd out of the brush, counted the animals, and drove them to the railhead. The number he arrived with was a measure of his success. Today, however, he measures his progress by the improvement in performance of each succeeding generation.

This book is an account of the trends and happenings traced through the beginning of cattle in America from the time they began having economic value to the modern-day age, when measuring devices are used to determine value and to predict genetic response to selection.

Today Polled Herefords graze the grassy plains of North America, the tropics of Central and South America, the wind-swept islands of South Chile, and the variable pastures of South Africa. There is not a land around the world where Polled Herefords have not adapted and helped improve the genetic base of indigenous cattle. From Warren Gammon's humble idea, bolstered by the faith and determination of his many fellow men of that day and those who followed, has evolved a movement that has had a great impact on the beef industry, and all mankind has become the beneficiary. Polled Hereford cattle, along with the people who breed them, have emerged as the most powerful force affecting genetic change in the beef cattle worldwide.

Contents

Acknowledgments

Acknowledgment is due all those who produce Herefords, whether they be Polled or horned proponents. It is particularly noteworthy that these Hereford breeders could endure for so long during a period that was so trying for both groups. The fact that they did endure, however, is what has produced today a relationship of mutual respect, understanding and friendly competition in the marketplace.

In addition, these few words are dedicated to and in acknowledgment of the staff members and employees of the American Polled Hereford Association who hitched their wagon to the same star.

Special recognition is given to members of the staff of the American Polled Hereford Association who served the breeders of Polled Herefords during those perilous times when its very existence was being threatened by dominant competitors and those who scoffed at the future of Polled Herefords.

There were employees who dedicated their time and effort toward serving the breed and providing services for the breeders when there was no assurance that their pay check would be waiting for them at the end of the month. In those later days, from

1960 to 1970, through the various stages of the single joint
agreement with the American Hereford Association, and the
development of the single joint certificate program, the dedication of certain staff members brought about the success of this
effort, and assured the fulfillment of the commitments that were
made by the American Polled Hereford Association's Board of
Directors.

It was during this period of time that the staff team made up of
Jim Harris, Temple Wells, Finis McFarland and Marcine Ragan,
performed seemingly impossible tasks. It was the unswerving
faith and loyalty of these four that made it possible to endure five
years under the burdensome yoke of the single joint certificate
program.

The *Polled Hereford World* magazine was a privately-owned
publication at that time, but the loyalty of its staff members,
namely Frank Farley, Jr., John Hall and Marilyn Sponsler, provided a line of communication between the association and its
breeders that made it possible to develop a closely knit organization that more clearly understood the problems and the challenges that the breed was facing. Up until this time and through
the contract period, staff members of the APHA and the *Polled
Hereford World* magazine found themselves forced to perform as
best they could under circumstances over which they had little
control.

During the period, however, from 1967 to 1970, which could
best be described as the period of negotiation, termination, and
transition, the preparation and capability of the staff of the
association made it possible for the Board of Directors to
negotiate from a position of flexibility. It was primarily staff talent
that gave the Board of Directors alternatives rather than be forced
to accept ultimatums.

For the five year period of time that the American Hereford
Association had control of the record keeping system, the

American Polled Hereford Association's staff developed the capability of reconstructing records in a complete form to enable the association to resume registration without skips in ancestry. This job which was nothing short of a miracle was done by Ken Harwell, for the APHA, Ed McLoud, Harold Kubler, and Bill Burns of Systronics, Inc. Included in this unusual accomplishment should be the names of Finis McFarland, Marcine Ragan, Temple Wells and Jim Harris.

Special recognition and mention also should be given to the loyal members of the APHA field staff, some of whom are no longer with the association, but who, by working closely with breeders in the field and making them aware of the problems that existed brought about a better understanding between breeder and staff. Those staff members were: Paul Aycock, Harold Schroeder, Charles Talley, Herb Brandner, Ralph Cook, Gene Kuykendall, Mack Patton, Eddie Sims, Charles Dooley, Bill Perry, Sam Wells, Kenneth Vaughan and Shelby Kahrs.

The APHA stands today as one of the most effective of modern breed associations. A close observation will reveal the finger prints of thousands. Most clearly are the imprints of those who depended solely upon Polled Herefords for their livelihood and their existence and the future of their families.

And last, a special spot reserved for one Matthew J. Heartney, Jr., who has the longest tenure of any in service to Polled Hereford breeders. First, starting as an odd-job boy in the shop of his father, who printed some of the first promotional material for Polled Hereford breeders. Later, after receiving his law degree, he became the legal counsel for the APHA and has served in that capacity for more than thirty years. Were it not for his wise counsel and guidance from behind the scenes it is likely that many costly mistakes could have been made and that the wonderful progress that Polled Hereford breeders and their association has made in the past years would have been impaired.

ORVILLE K. SWEET

Appreciation

It was only because of the great spirit and generous help of the staff of the American Polled Hereford Association and breeders of Polled Herefords that this book has become a reality. A great deal of the information about personalities and history was made available by pioneer breeders and their families. Special thanks is due Temple Wells, Administrative Secretary for the American Polled Hereford Association, who worked tirelessly in assisting with research and typing, and Don O. Smith, former Art Director for the *Polled Hereford World*, whose many fine illustrations appear throughout the book.

Breeders who have given special help and whose names add honesty and validity to this work are Walter and Frances Lewis, Larned, Kansas, and Jim and Fay Gill, Coleman, Texas.

I am also grateful to Doug Petty, Payson Lowell and the staff of The Lowell Press for their many helpful suggestions; and to Doris Morris for her editorial assistance.

KANSAS CITY MISSOURI   ORVILLE K. SWEET

1. The Beginning of Beef

The Legend of the Longhorn
The Original Master Breeders
The Whiteface Performance Test

The Legend of the Longhorn

The first "Mexican" cow splashed her way across the Rio Grande nearly 450 years ago, paused to swat a persistent fly, and began grazing on the north banks. She was not an animal of proud pedigree, but she was the forerunner of today's modern beef cattle. Her color, gauntness, and perversity were historic. She was a product of generations of hard environment. She was the first of millions that were to fatten on the grasses of the border and eventually to migrate to the north.

She came to what is now the United States with the missionaries from Mexico, but the thefts by the Indians, our red brothers, the adversity of the elements, and general discouragement forced the Spanish missionaries and soldiers to abandon the challenge and return about 1693 to the protection of a more "civilized" land. The Longhorn cow remained to become the founder of an empire. She matched wits with the wilderness, met claw and fang with horns and cowsense. When the Spaniards returned twenty-three years later, Longhorn cattle were grazing on the grassy plains of what is now northeast Texas. Since that day Texas has never been without cattle.

The Longhorn cow fills a unique place in the history of American animal husbandry. She became the genetic base of this country's fifty-five million commercial beef cows. She has provided hardiness, immunity, and adaptability, the like of which has not been possessed by any other breed or species of domesticated livestock. She fled to the nearest pool to escape the bite of the heel fly. She fought off the wolves by night and outran the Indians by day. It is to the hardiness of the Longhorn that imported English breeds owe their successful performance on American ranches. It is doubtful that any of the offspring of the English or Scottish breeds, even with their royal ancestry, could have survived the raw, wilderness environment, the infestations of parasites, and disease without a measure of the Longhorn toughness with which to integrate.

Long on hardiness, horns, and longevity, this tough, hardy breed came on the American scene and survived. Their long legs allowed them to travel great distances to railheads and to traverse over mountainous terrain. Their sinewy, tough muscles helped carry them to the sparsely located water holes throughout the

A Longhorn steer.

range area. They were described by screen star James Stewart in the motion picture *The Rare Breed* as "meatless, milkless, and murderous." But in spite of their low quality, they were adequate for the needs of a young, growing country whose taste had not developed to the point of demanding a high-quality steak, as it does today. As living standards improved and the desire for better steaks developed, the stately, aristocratic Shorthorn from Scotland came on the American scene. Expertly promoted, the two breeds became a natural cross. At that time it would have been too revolutionary to talk about removing the horns from the Longhorn steer, but their promoters exploited the Shorthorns in the most effective way, by claiming to shorten the murderous weapons on the head of the Longhorns.

A Shorthorn-Longhorn cross succeeded in improving the milk production of the Longhorn cow, increasing the weaning weight of her calf and then improving fleshing quality. It was a perfect example of one breed complementing another to overcome the traits in which each was deficient. The cross was designed primarily to improve the Longhorn in the basic economic traits—early maturity, mothering ability, and not the least, disposition.

When we trace the cattle history of the United States, we find that the commercial breeder has practiced crossbreeding and exploited the benefits of heterosis from the beginning. But at first the lowly, undignified Hereford and the repulsive Angus, both in the minority, were simply observers of this change on the American beef-cattle scene. The Shorthorn breed completely dominated the livestock shows of America, winning almost every carcass and steer contest at the annual International Livestock Show in Chicago.

At the turn of the twentieth century the grand-champion steers at Chicago were still weighing more than twenty-five hundred pounds and averaging four to five years in age. With the advent of

A typical grand champion Hereford steer at the turn of the century.

United States Department of Agriculture (USDA) grading standards in the early 1920's and the modern butcher's block, it was
evident that carcass size of steers must be reduced because of the
so-called changing taste of the American consumer and the
alleged demands of the American housewife for smaller cuts. The
need became more evident for an earlier-maturing animal that
would produce a high-quality carcass at an early age. Thus the
Hereford was lofted into prominence on the beef-cattle scene.

In 1880 an alert and aggressive group of Hereford breeders met
in Chicago and determined that the public should have an
opportunity to view the best the Hereford breed had to offer.
Realizing the importance and the impact of collective effort and
promotion, these early Hereford breeders banded together and
pledged themselves to castrate at least five hundred calves so that
they might bring the best they had to offer to the International
Livestock Show and combat the long, successful history of the

Shorthorns in the international steer show. Thus was born the American Hereford Association for the purpose of promoting the Hereford in the United States.

The first major goal of the Hereford breeders was to sell Hereford bulls for crossing on the predominantly Shorthorn cows throughout the commercial cattle industry. Herefords were promoted on their ability to cross with the Shorthorn cow herds of America and produce improvement in earlier maturity, smoother flesh and hardiness, and the ability to withstand the severe weather and the long, cold winters unprotected by barns or cover in the West. It is interesting to note at this time the beginning of the decline of the second great breed, Shorthorns, as a dominant factor in commercial cow herds in the United States—first, the rise and dwindling away of the Longhorn and then the demise of the Shorthorn.

Both the Shorthorn and the Hereford came to prominence through the practice of crossbreeding with the established cow herds of predecessor breeds throughout the commercial areas of America. Both breeds had something in the way of improvement to offer over the established cow herds, but at the same time both breeds rose to prominence while the hybrid vigor resulting from crossing with another breed was exploited. Over a period of approximately twenty years the cow herds of America became predominantly whiteface.

When the rate of progress declines, the cowman naturally looks elsewhere for genetic material to build a new beast that will result in increased improvement in the basic economic traits. Crossing does not always give him that new impetus. When, however, the registered breeder fails to keep the level of performance in his registered herd above that of the commercial herd, then the commercial man will look elsewhere or to other breeds. As one travels through the cattle-producing areas of this country today, he sees the pastures populated with multicolored cattle. The spotted and varicolored commercial herds throughout the cow-producing areas are the result of frustrated attempts on the part of the commercial man to improve his level of production. He obviously has not found it possible to do so by using bulls from the breed common to his cow herd. For that reason there has been a constant succession of strange breeds streaking through the American beef-cattle scene, each of them having an influence to some degree, most of them, however, claiming to have something different and greater than the others to offer the cattleman.

Few industries without facts and guidelines to follow can exist without being swept along by the winds of change and are thus affected by fads and fancies brought on by strong personalities. Personal preferences of individuals with fat promotional budgets will have a great influence. The beef industry has been no exception. Color fads, extremes in type and conformation, and even pedigree fads have come and gone. Fortunately, the bovine gene package has protected the species from the crude methods of the exploiter who promotes for profit only. A breeder once said: "I would rather have hair than bone any day. You can make them look better, even if they ain't." Another breeder made this statement to his manager, "It doesn't really make any difference if we need a new herd bull, we can continue to fool 'em with what we have." These statements and attitudes were not necessarily dishonest but were simply among the cattlemen's bag of marketing tools that he could delve into for use whenever the situation called for it. A veteran breeder summed up the situation on one occasion when he said, "I sure do hope the next popular bull is a good one."

The cattleman is a realist in that he has to pay the bills monthly and the mortgage on his ranch each year. He finds it necessary to use every tool and opportunity to get the utmost for each animal that he sells, but in so doing, he slows the rate at which he can progress genetically.

The Original Master Breeders

The English and Scottish cattlemen were early master breeders. They were keen observers of animals and were master merchandisers. Because most of America's major breeds originated in their countries, they were able to promote some ideas and practices that had a great influence on early American cattlemen. They were the first to develop techniques and trade secrets that gave them a dominant position in the world in seed-stock marketing. In the early days with each exportation to North or South America, they arranged for one of their own countrymen to accompany the cattle and in most cases to remain with them as the main attendant manager or advisor-consultant. Although they gave Americans many useful suggestions, based on their keen observations and instinct for animal care and propagation, Americans also gullibly accepted many ideas and practices that were not necessarily constructive.

For example, a common sight at cattle shows early in the twentieth century was the Scottish herdsman sitting on his tack box, peeling and chopping turnips or preparing some other feed additive assumed to have some mysterious ingredient that ensured success in the fattening and fitting process. Another good example was the use of foster mothers as nurse cows to put extra bloom and flesh on calves. In some parts of the world the practice had gone so far as to continue to supplement an animal's feed with milk from a bucket when he was too large to kneel and nurse.

It was a common philosophy among those early-day herdsmen that selecting and mating cattle was surrounded with an aura of mystery not understandable to the layman or neophyte: "One had to have a special insight or a feel for the art in order to become a seed-stock producer." They claimed unto themselves a special talent that could not be developed through exposure and experience but only through inheritance or by "the laying on of hands" by the priestly tribe of herdsmen. This philosophy with its hocus-pocus prevailed in the minds of early American cattlemen and delayed for many decades the introduction of thought and reason in the cattle business.

The Whiteface Performance Test

I find that the great thing in this world is not so much where we stand, as in what direction we are moving. We must sail sometimes with the wind and sometimes against it, but we must sail and not drift nor lie at anchor. This bit of wisdom originating with Oliver Wendell Holmes more than a hundred years ago is apt for the Polled Hereford breed.

The breed has been constantly changing to meet changing consumer demands. First the change from slow-maturing animals of large size to an earlier-maturing, faster-growing individual. The resulting animal yielded smaller, more desirable cuts for America's family dinner table which at one time was set for an average of seven to eight people with hearty appetites but diminished to an average of three and a half persons.

The initial shift toward efficiency, of course, was the elimination of horns and the costly brutal process of dehorning.

Beyond these two dramatic adaptations, whiteface cattle, with or without horns, had a great deal in common. The most important common trait, and that which determined a breed's survival, was adaptation to environment, particularly weather extremes and available feed supplies.

Breeds of beef cattle, like other products of industry, can be in demand and thrive for a short period of time on fat promotional budgets, but the test of time is brutal and fatal to those that fail to perform. When a breed attempts to invade the American beef-cattle domain, it is challenging those breeds that are native to the wide range of adverse weather and various feeding conditions. This is an experience that only a few breeds can survive. Throughout this century many breeds have made an attempt, but only a few have succeeded. The performance test on the western range country of the United States has proved too much for most of the "hot-house" breeds of Europe that were pampered and cared for as backyard cattle. The average cow in the United States is expected to endure temperature ranges from $-30°$ F. to $110°$ F. In the early days she had to stand belly-deep in snow with her tail tucked into a fifty-mile-an-hour wind at below-zero temperatures with an inch of ice on her back and wait for days for hay or feed.

It is possible for breeds to multiply and thrive for a few years until a very severe winter comes along. It is the breed that walks out of the snowbanks in early spring that passes the test and claims the dominance of the range. In the early days the cow was expected not only to survive the winter storm but to meet the spring thaw and green grass with a healthy, thrifty new calf that would bounce up within a few minutes after birth, nurse a few ounces of milk, and be strong enough to outrun a coyote.

The three-way test for any profitable cow is to have a calf every twelve months unassisted, give enough milk to raise it, and provide it with enough inherent growth potential to weigh heavy for a profit.

Any breed that fails to meet all three of these test factors will become a passing fad. Fertility and reproduction under adverse conditions is the most important breed characteristic in beef production. The value and essence of the Hereford breed has been greatly enhanced by this breed characteristic. It has been so from the beginning.

Proof of a good cow.

The Test of Time

As time goes, Herefords are a fairly recent import to the shores of America. Only a bare handful, less than three hundred head, had been imported to the United States prior to eighty years ago. It was within the short span of about ten years, however, that the whiteface proved its adaptability and resistance to the breed began to subside. It was during the period of the beef industry's rapid expansion in America that the Hereford made its entry and dramatic surge across the plains and mountains of the central and western parts of the United States. Since that time the whiteface has reigned supreme. Eight decades is not a long time in the history of man but it is a significant period in the realm of improved seedstock. No other breed has grown so fast, adapted so well and remained in dominance so long as has the whiteface cow. She has adapted so well to all sections of America and Canada, with their extremes in range conditions, that her performance test for world-wide conditions is now complete.

The Test of Climate

"It was during the severe winters of 1889 and 1890, I lost about 30,000 cattle, or nearly 65 percent of my entire herd," wrote the late Governor John Sparks of Reno, Nevada. "The Herefords at that time constituted about 40 percent of the herd, and I found that of the entire number surviving the second winter at least 90 percent were whitefaces, showing conclusively their superior constitution."

"One of the worst storms in history of this country began February 12, 1912," wrote A. M. Dunn, Clayton, New Mexico. "Snow fell several days, followed by sleet which left a crust on the snow and made it impossible for the cattle on the range to get at the grass. When the blizzard began, Union County was covered with cattle, mostly grades, very much mixed as to breeding. When grass came in the spring the whitefaces were about all that were left."

There is a legend that has been handed down through generations of cattlemen about survival traits and characteristics of cattle. During the cold blast of winter's northwinds, a common trait of all livestock is to turn their backs to the wind and begin to drift downwind. Small groups gather into herds and herds drift into larger droves. They all move slowly downwind until they meet a fence, canyon, river or some other obstruction that stops their southward movement. Here the pushing and shoving begins, each maneuvering for a warmer spot as they strive to avoid the cold wind. There is piling up, tramping and smothering. Great losses usually occur as the result of crowding and suffocation. Legend has it that the Hereford, rather than turning its tail into the wind may drift, but will ultimately turn its head into the wind, thus avoiding the tendency to over-crowd.

Research shows that the Hereford has a slightly thicker hide which gives a bit more insulation from the severe cold.

Although the whiteface cow may not give quite as much milk during nursing season, this trait becomes an asset as she goes into the winter with a slightly bit more fat for self-protection and preservation.

The testimony of pioneer cattlemen includes examples of the constitutional ruggedness of the whiteface that has been demonstrated numerous times.

As recent as the winter of 1974-75, entire herds were wiped out in Wyoming, Nebraska, Iowa, North Dakota and South Dakota. Some farmers and ranchers lost their entire calf crops. Those who fared the best were the ranchers with whiteface cattle.

The Test of Hardiness and Prolificy

In addition to the ability to survive adverse climate conditions the most convincing performance trait according to commercial cattlemen is the cow's ability to bring in a live calf every year. Without a calf to sell to pay the maintenance cost for keeping the cow, the rancher is out of business. Of all the breeds to challenge the Hereford cow on her native range, none has been able to beat her at her profession of producing a high percentage of live calves each year. The weight of her calf may vary from year to year as available feed varies, but she can beat all competition in conceiving and raising a live calf.

The fertility of her helpmate, the whiteface bulls, is equally respected. Polled Hereford bulls are unexcelled in rustling over large pastures and range land, hustling their own feed and servicing the herd as the cows cycle and come in heat.

The famous Matador Land and Cattle Company for many years was ably managed by Murdo Mackenzie who cited their prolificy as one reason for his partiality to whiteface bulls.

The ranges of the Matador Ranches were large and grass was sparse. With whiteface bulls they usually got an 80 percent calf crop, but when they tried using another breed, the percentage dropped to 40 percent. A return to the use of Hereford bulls quickly brought about a corresponding increase in the calf crop.

As early as 1915 the United States Department of Agriculture issued a farmer's bulletin which stated, "As rustlers, Herefords are surpassed by no breed of beef cattle and they excel the Shorthorn in this respect. They have been recognized as a breed which responds readily to favorable environment, as well as being able to thrive under adverse conditions where other breeds would not do well. On scant pastures and on open range where waterholes are far apart, the Hereford has shown merit. The bulls are active, vigorous, prepotent and very sure breeders." (USDA Farmer's Bulletin No. 62, Jan. 21, 1915.)

The Market Test

The most crucial test for any product is its value on the competitive market. The moment of truth for a breed is when the feedlot operator gives his opinion at the market place. He bids his price based on his estimate of the efficiency with which that breed will convert feed to high quality, edible beef for him to sell to the packer at a premium.

There are those who will question the value of color saying, "Color is of no value when the hide is off, one can't even identify the breed at this point." This is true, but the main consideration is whether or not color is of any value as an indicator of performance and efficiency while the animal is still alive.

Great sums of money have been spent on research in an attempt to determine the value of color. In summarizing all that has been learned to date, we must conclude that minor differences in color are of little significance within families or breeds. However, there are great differences between breeds when comparing specific traits.

The size of a newborn calf is a highly inherited trait which varies between breeds. A calf being too large is the greatest cause of calf losses at birth. On the other hand, calves that are smaller than average at birth are less likely to grow as fast or have as great a mature size. Average birth weight is associated with breeds, therefore, one may correlate calving difficulty and growth rate with breed colors.

The color code system has become an effective tool throughout the industry in estimating performance. A recent survey of the huge feedlots in the high plains area of Texas, Colorado, and Kansas revealed that 70-80 percent of all the calves in the feedlots were whitefaces. In this day of performance awareness, feedlot operators and feeder cattle buyers have shown a decided preference for whiteface cattle because of their reliable performance in the feedlot and desirable carcasses.

The demand for Polled Hereford bulls has exceeded the production by seed stock producers since the breed's inception 75 years ago. Because of their superior performance in competitive feed tests, their natural rustling ability and eagerness to breed, the demand for Polled Hereford bulls has outstripped the demand for any other breed. During the onslaught of the exotics, the Polled Hereford bull market remained strong. Because of the failure of the continental or exotic breeds to perform profitably in the feedlot, many commercial breeders are returning to use of the reliable breeds. Polled Herefords are enjoying a preferred position during this switch back to the basic breeds.

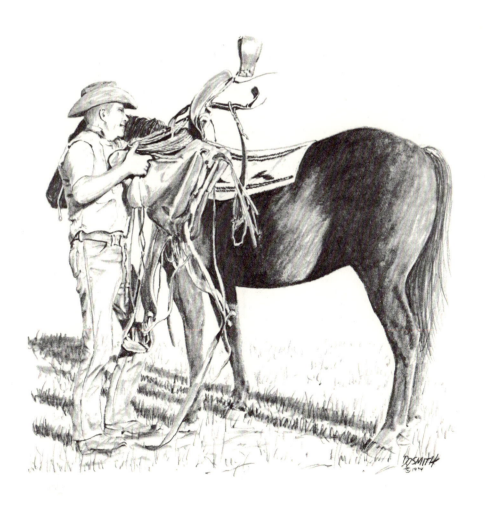

2. Early Herefords in England

The Breed's First Breeders
First Herefords in the Homeland

The Breed's First Breeders

The changes we observe in type and performance of cattle are the result of what man has done to them. Improved breeds are man made. One English historian terms cattle as either the result of *natural* or *artificial* selection. He applies the latter term to the means used by man to breed animals to conform to his ideal. Man projects a specific ideal that he has mentally developed and strives through selection and trial and error to obtain his goal.

It was a noble effort on the part of the Benjamin Tomkins family to undertake the task of selective improvement in the 18th century. With only the crudest of tools and limited knowledge, it must have been discouraging for them indeed. Unlike poultry, swine, or rabbits that can reproduce rapidly, cattle require at least four years to completely manifest genetic improvement from a single mating. In his normal lifespan, if he devotes his entire adult life, a man may see the results of six to ten generations of his selective matings. It is only when dealing in great numbers that rapid progress can be made.

Benjamin Tomkins, the elder, was the first improver of Herefords, while his son, Benjamin the younger, as he was called by some historians, is credited to have made the greatest contribution toward establishing the permanent identity of the breed in

Benjamin Tomkins, the younger.

Herefordshire.

The lives of these two devoted Hereford breeders spanned from 1714 to 1815.

A most unusual story involves the elder Tomkins and the earliest instance of cattle being named in a document of record. The cow, "Silver with her calf" was listed in the will of his father, Richard Tomkins, in 1720, and was bequeathed to the son, Benjamin Tomkins, the elder, who was only six years of age at the time. A foundation family of Herefords resulted from the cow Silver and it may have been unusual insight on the part of the father to entrust to Benjamin, the fourth of six sons, the cow that later became one of the founding matrons of a breed.

Color was not considered as being among the prime traits as was evidenced by the first bulls selected for use by the Tomkinses. Some were of mottled face, some solid red or grey. The trait upon which they placed greatest emphasis was ease of fleshing and growth.

Silver cow.

The Silver cow was described as having a line of white down her back. Nearly twenty years later a bull was produced by the younger Tomkins called "Silver Bull," tracing to the Silver cow and likewise having a stripe down his back. It may be said today that many of the cattle with this characteristic descended from the original old Silver cow of 1720. Tomkins' herd contained not only the Silver, but Pidgeon and Mottle families and in fact for many years these family names were used to distinguish the different lines. Silver referred to a red cow with whiteface and streaked back; Pidgeon, a grey cow; and Mottle, a dark red cow with spotted face.

By keenly observing a pasture of registered whiteface cows even today, one can discern likenesses of the foundation cows. The Mottles and Silvers can be recognized by similar color markings. It is fortunate for today's Polled Hereford breeders that the original breeders of Herefords so carefully selected the traits that had the highest market value. We are likewise indebted to them and later generations of breeders who so jealously guarded those traits while striving for perfection in others.

First Herefords in the Homeland

Although some early writers have attempted to give an "accurate and complete" account of the origin of the Hereford, on searching the many recorded stories one is a bit confused and less certain about the first few generations.

Hereford history begins with a story of nondescript cattle identified in Herefordshire, England, in the early 1600's. Probably the most reliable account of the history of Herefords is by McDonald and Sinclair, first published in 1886. They briefly refer to John Speed, a London writer, in 1627, who stated, "The climate of Herefordshire is most healthful, and the soyle so fertile for corne and cattle, that no place in England yieldeth more or better conditioned."

I have been unable to find any special historical allusion to the Hereford breed before or during the seventeenth century. This omission is explained by the circumstance that it was not until about the end of the eighteenth century, when British agriculture raised up its own chroniclers, that breeds of livestock attracted much notice. Since the advent of the agricultural historian, however, this variety has received a good deal of attention.

Thomas Duckham, who was for many years the editor of the Hereford Herd Book, in an address to the students of the Royal

Agriculture College stated that little was known of the breed and history revealed very little prior to the beginning of the Smithfield Fat Stock Show in 1799.

In a written account in 1788, William Marshall revealed, "These several breeds I conceive to have sprung from the same stock. Their colour apart, they perfectly resemble the wild cattle which are still preserved in Chillingham Park, and it appears to me that the different breeds above noticed are varieties arising from soils and management of the native breed of this island. The black mountain breeds of Scotland and Wales appear to me evidently to be from the same race, agreeing in everything but colour with the red breeds that are here adduced."

Marshall further described the cattle of Herefordshire as being very similar to those in Devonshire in frame, color and horn, but were not so clean of head and larger in size.

It was thought that the cattle of Herefordshire were the first breed of cattle on the Island and resembled many others at the time, but were outstanding in their size.

They were described as athletic and good travelers, having a form nearly perfect for draft stock. They fattened well and especially were impressive as three-year-old fat heifers.

Terms used to describe cattle over the years have evolved and gone through several translations. Even in the past decade there has been a revolutionary change in terminology to describe a more desirable animal. It is interesting to note the very careful and specific way Marshall, a respected English writer, described the Hereford in 1789:

"The general appearance full of health and vigour, and wearing the marks of sufficient maturity—provincially 'oxey,' not steerish or still in too growing a state to fat. The countenance pleasant, cheerful, open; the forehead broad; the eye full and lively; the horns bright, tapering, and spreading; the head small; the chap clean; the neck long and tapering; the chest deep; the bosom

Early Herefordshire cattle.
(Reproduced from Volume I, Hereford Herd Book.)

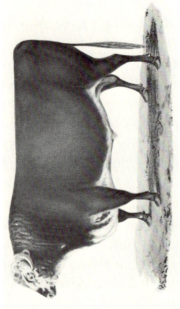

Wellington (4), born 1808.

Brockswood (485), born 1843.

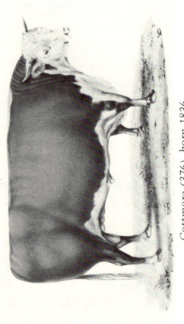

Cottmore (376), born 1836.

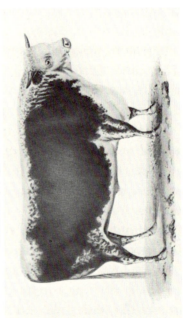

Victory (33), born 1839.

broad and projecting forward; the shoulder-bone thin, flat, no way protuberant in bone, but full and mellow in flesh; the chine full; the loin broad; the hips standing wide and level with the spine; the quarters long and wide at the nache; the rump even with the general level of the back, not drooping, nor standing high and sharp above the quarters; the tail slender and neatly haired; the barrel round and roomy, the carcase throughout being deep and well spread; the ribs broad, standing close, and flat on the outer surface, forming a smooth, even barrel, the hindmost large and of full length; the round bone small, snug, not prominent; the thigh clean and regularly tapering; the legs upright and short; the bone below the knee and hough small; the feet of a middle size; the cod and twist round and full; the flank large; the flesh everywhere mellow, soft, and yielding pleasantly to the touch, especially on the chine, the shoulders, and the ribs; the hide mellow and supple, of a middle thickness, and loose on the nache and huckle, the coat neatly haired, bright and silky; its colour a middle red, with a bald face, the last being esteemed characteristic of the true Herefordshire breed."

This then, is the picture of a typical Hereford as the breed existed about a hundred and eighty-six years ago. It is so complete that it is not to be wondered that later writers acknowledged they were unable to improve upon the description, which they accepted as the best that could be given. It is well in passing to emphasize the fact that during the closing decade of the eighteenth century the Hereford was in color a middle red, and that a "bald face" was then a characteristic of the true breed.

One should not lose sight of the fact that domesticated cattle were mostly triple-purpose until the twentieth century. They were raised first for draft or work, second for dairy, and third for beef when they had outlived their usefulness for the first two purposes. This multiplicity of traits, as described by Marshall, made selection somewhat more complex, but at least the priorities were well understood.

The cattleman of today would be reluctant to go into such great detail. He would be concerned that he would lose sight of the major economic traits and the need to place heavy emphasis on these important traits to make more rapid change.

Early Matings—Source of Germ Plasm

A review of all the known circumstances connected with the origin and early development of the Hereford breed seems to establish the fact that it was founded on a variety of the aboriginal cattle of the country—the same type from which the Devon and Sussex breeds derived—and that the original color was probably a solid red. At an early period the Welsh white cattle—which were not only different in color but larger in size than the local variety, and probably containing some foreign blood—were introduced and mixed with the stock of Herefordshire, giving a tendency to white markings, and enlarging the frames of the native breed.

A very eminent scientist and author, a Professor Dawkins, did considerable research on the early settler of England and the original animals they found there. It was his studied opinion that during the Roman times only the small cattle of the Welsh and Scotch type were present. They belonged to the strain, Bos Longifrom. At about the time of the English Conquest more cattle were introduced from Holland, of the Bos Primigenius strain. They were larger cattle and probably the first whitefaces to be introduced.

The question of the origin of the whiteface and its dominance has always been an elusive one and probably will never be answered for certain. The most logical explanation is that given in about 1800 by an old gentleman farmer whose identity is about as obscure as the answer he attempted to give: "The Hereford breed originally were all red cattle—red faces, etc., when a bull brought in from another part, with a whiteface, proving a good getter, was the forerunner of this most excellent breed which I am inclined to think, if not the best, are as good graziers' beasts as any in England; obviously the red cattle of Herefordshire were much improved by the cross."

We shall lay the question of color, which Charles Darwin calls the most fleeting of characters, to rest with the basic assumption that color is important if it can be interpreted as a color code for performance.

The Polled Gene

It is interesting to note that most historians, in describing the early day cattle of Herefordshire or in referring to typical Herefords, include a description of the horns. It can be concluded from the general reference to horns that polled or natural hornless cattle either were not preferred or existed as a rarity. It would be extraordinary, however, if the polled gene did not exist in the many varieties of nondescript cattle the historians record in the Herefordshire cattle. Although other natural hornless breeds did not have their birthplace in Herefordshire, England was the origin of most polled breeds, and all of them trace to indigenous varieties that were developed from the sixteenth to the nineteenth centuries.

It is reasonable to assume that the polled gene lay submerged and dormant in the populations of cattle being developed and was a victim of prejudice during a time when breeders selected away from it while culling all those that emerged born without horns. Although genetic mutation has been the most popular explanation for the existence of the polled gene, we must recognize that mutation or crossbreeding was not a necessary conclusion, since the polled gene has existed in cattle since the days of the Egyptian Pharaohs. Inscribed pictures of polled cattle have been found on the walls of tombs and in other archaeological findings in Egypt.

It is not my purpose here to trace in detail all the lines

established in early Hereford history, but to credit those original breeders who were considered seed-stock producers. A seed-stock producer is one who, through a system of selection, endeavors to fix traits of economic importance and establish a breed identity.

Traits of Economic Importance

It is also interesting to see how traits have varied over the years in economic importance as changing times and needs have dictated. It was well known that the production of beef or milk was not then the sole purpose in cattle-breeding nor were these traits considered first in priority. Usefulness for the purpose of labor in the field was generally regarded as being more important. Cattle were, in many parts, used chiefly for draft, and it was only after they had outlived their usefulness as work animals were they slaughtered for beef.

Although milk production was a secondary trait in those early years, it is interesting to note that in time the Hereford was used for dairy purposes. Many dairies throughout England used the Hereford cow predominantly for milk production until the last Hereford dairy herd was dispersed about 1947.

Breeding for Beef

Some early selections had taken place in the formative stages of the Hereford breed, but it is unlikely that selective breeding began in a serious vein until the middle of the eighteenth century. There is ample evidence that it began first with Hereford cattle.

The science of selecting certain traits or characters and applying intensive pressure on these traits through several generations is known as genetic engineering.

It may be the most intriguing but elusive of sciences, when applied to animals. The manipulation of prepotent material that you can neither see, feel nor measure in any way has to be one of the most challenging involvements of man. The great horse trainer, Sonny "Jim" Fitzsimmons, put it most simply, "It is not what you see that matters."

Selection, followed by mating the "best to the best," has long been the principal tool through which man has attempted to bring about improvement of domestic animals. Since the practice of selection was begun, man has chosen certain animals to keep for breeding while consigning the balance for slaughter and to be eaten.

Until very recent times, the general principle that "like begats like" was the only recognized law of heredity. This principle has been effective over the long period in modifying animal types, as one may assume when comparing types of today with those of a few years ago.

To Robert Bakewell, the English patriarch of the eighteenth century, much credit must be given. A study of his theories and concepts of selective breeding should mark him as the first researcher pointing the way to livestock improvement. A hundred years before Mendal's laws became known, Bakewell's practice of progeny testing introduced a new age in animal husbandry. His arrangement of lending rams to farmers and subsequently measuring the progeny for the purpose of making comparisons was an "idea before its time." His intensely practical ideas and emphasis on the utility value and market requirements was not to be applied by seed-stock producers in a serious way for another two hundred years. Bakewell's imaginative insight may be compared to that of Leonardo da Vinci whose fifteenth century drawings of rockets and flying machines preceded the invention of the airplane by more than four hundred years.

3. America's First Beef Breed

The Origin and Growth of Polled Herefords

Polled Herefords represent the development of an idea—an idea born in the minds of a small number of middle western Hereford breeders in the late 1890's who realized that it was both possible and practical to develop "modern Herefords minus horns."

These breeders were motivated by the promising prospect of maintaining the outstanding beef-producing characteristics of Herefords, but with the added desirable trait of being naturally hornless. They planted the seed from which was to grow a new giant on the American beef-cattle scene.

The Polled Hereford of today is a modern, practical breed of cattle that has experienced such widespread acceptance and desirability that its popularity is magnified by an unparalleled growth record. The Polled Hereford has often been called the product of necessity, a second stage in the evolution of horned Herefords.

The Hereford breed was founded in the mid-eighteenth century by the farmers of Hereford County, England. The first recorded Herefords were imported to the United States in 1816 or 1817 by the famed Kentucky statesman Henry Clay. Among

Polled Hereford founder Warren Gammon, Des Moines, Iowa.

the horned Herefords an occasional calf did not develop horns. This unexpected departure from the parents' characteristics is known as a mutation. Such cattle soon came to be called "polled," which means "naturally hornless."

History records the existence of polled, or naturally hornless, beef cattle since the first descriptions of civilization. Etchings and carvings in the Pyramids show an occasional polled animal, dating more than four thousand years ago. Also, over the years it has been established that the polled gene is dominant over the horned trait, so that more often than not polled offspring result from mating polled and horned animals. A number of pioneer American breeders developed breeding programs involving polled beef cattle.

It was not until 1898, however, that the first serious breeding program was started with the goal of producing Polled Herefords with the use of registered Herefords exclusively. This program represents the origin of Polled Herefords. The idea was conceived by an Iowa lawyer, Warren Gammon, of Des Moines, after he saw some polled cattle on exhibition at the Trans-Mississippi International Exposition in Omaha, Nebraska, in 1898.

Gammon did extensive reading on the subject, and what particularly affected his decision was what he read in Charles Darwin's works *Origin of the Species* and *Plants and Animals Under Domestication.* In reality, it was a graduate paper written by his son, Bert, that first sparked the elder Gammon's interest in studying Darwin's theory. Darwin discussed at length the matters of mutations and variations and how they could be made permanent by systematic matings. He concluded that in every variety of plant and animal life all sorts of curious "freaks" of nature were constantly occurring.

Gammon seized upon the idea of locating a few purebred Herefords that had failed to develop horns and from that start to develop a naturally hornless strain of whiteface cattle that would be purebred and eligible for registration. In 1900 he sent inquiries to the 2,500 members of the American Hereford Association, attempting to find some naturally hornless purebred Herefords. From the some 1,500 replies he received, Gammon found and bought four bulls and ten cows. Two cows were barren, and one bull was eliminated and so it was from the remaining eleven animals that Gammon established the Polled Hereford

LUTHER BURBANK
SANTA ROSA, CALIFORNIA
U. S. A.

September
Seventh
1 9 1 8

Mr. Warren Gammon,
Des Moines, Iowa.

My dear Mr. Gammon:

Your esteemed letter of September 3d just received
and I hasten to reply though not as fully as I
would like as my time is priceless beyond expression.

The work which you have done in the natural de-horning
of stock cannot be counted by millions of dollars,
its value to the world must be counted in billions.
I have always thought how foolish it was for civilized
cattle to wear these remnants of barbarism. I once
knew a gentleman who was a neighbor in the East who
originated the Polled Jerseys.

While I have no words to express my appreciation of
your work with cattle I do not have quite as much
confidence in your botanical knowledge because
being a specialist like yourself I must disagree
with you in regard to bees though I cannot explain
it as I would like to you.
I inclose a little essay which I wrote several
years ago which may elucidate my views on the matter
as I have not time now to write more extensively.

Thanking you for your kindness in writing and giving
some of your experience, I remain one of your most
faithful admirers.

Sincerely yours

Luther Burbank

Excerpted letter from renowned botanist Luther Burbank.

breed registry. These original eleven Polled Herefords were registered in the American Hereford Association, but were not identified by their polled characteristic. Therefore, he formed the American Polled Hereford Cattle Club to maintain a separate record of purebred Polled Hereford registrations.

Thus, in 1901, the Polled Hereford breed came into being with eleven registrations on record. In 1907 the pioneer breeders of Polled Herefords incorporated their organization, with headquarters in Gammon's home in Des Moines. He served as executive secretary until 1911. He was succeeded by a son, B. O. (Bert) Gammon, who continued in this capacity until his retirement in 1946. D. W. Chittenden served as executive secretary until October 16, 1962, and L. J. Harris served temporarily until January 1, 1963, when I was appointed executive secretary.

In 1947 the association was renamed the American Polled Hereford Association (APHA), as it is known today, and in that year its headquarters were moved from Des Moines, Iowa, to Kansas City, Missouri. By 1957 the association had outgrown its leased quarters, and a successful building fund campaign was launched. On October 21, 1958, groundbreaking ceremonies signaled start of construction of the APHA's present headquarters building in Kansas City, on a three-and-one-half-acre tract in the south section of the city. The building was occupied on March 1, 1959, and dedication ceremonies were held on October 20 of that year.

Beginning with the original eleven Polled Herefords in 1901, the breed began to grow. Growth was necessarily slow at first but increased rapidly over the years to a total in January, 1973, of more than three million head. More than 60 per cent of this growth was recorded in the past decade.

In the early days of the breed Warren Gammon, the breed's founder, approached the American Hereford Association and

B. O. (Bert) Gammon.

made a plea for that association to include in its registry procedure a provision to indicate "polled" ancestry on all pedigrees. This would not have been a great problem except for the fact that the policymakers of the Hereford breed at that time refused to recognize any segment within the breed. Their general attitude was that Herefords were Herefords, with or without horns. Of course, underlying this philosophy was the suspicion that anything that was born without horns was not purebred and probably was the result of some neighbor's Red Poll bull sneaking through the fence at unattended times. If this was the case, however, the incident must have happened many times before because the polled gene, although infrequent, revealed itself throughout the Hereford breed.

A New Breed—Or the Evolution Of an Old One?

It obviously was not Gammon's intention to start a new breed or to open a separate breed registry. He found it necessary, however, in order to keep a record of polled ancestry. He initiated his own method of keeping a continuous lineage of animals introduced into the Polled Hereford breed. He indicated polled ancestry on each animal recorded and issued a three-generation pedigree.

Several attempts were made, in the long period of time from 1901 until 1947 when the Gammon family was founding and developing the breed, to have the American Hereford Association issue the registration certificates and designate in some way the naturally polled animals showing up in all pedigrees. The adamant stand by the policymakers of the American Hereford Association was maintained however: Herefords were all Herefords, and that there was to be no distinction shown, whether polled or horned.

It was not until 1951, at the initial meeting of the World Hereford Council in England, that questions were raised concerning the completeness of the records of the American Hereford

Association relating to horned and polled cattle. Delegates to that conference stated their concern that animals being purchased from American Hereford breeders might not be "pure Herefords" even though they were "crowned with the horns of purity." They might possibly have a taint in their pedigree by having an ancestor that was polled.

The Argentine delegate to the World Hereford Council asked Jack Turner of the American Hereford Association, "If I purchased an animal from your association, could you assure me that there was no polled ancestry in its pedigree?"

Turner answered, "No, not with certainty."

Upon his return to the United States, Turner instructed his staff members, and policy was established, that all animals without horns would be indicated with a "P" (polled) prefix before the number on the registration certificate.

Even though this practice was started in 1952, there still was the problem of one and one-half million Polled Herefords recorded in the American Hereford records before that time that did not have an identification or designation whether they were horned or polled. Gammon continued his original practice of requiring that all Polled Herefords first be registered in the American Hereford Association to prove their Hereford genesis and then be registered in the American Polled Hereford Association to prove their polledness. The American Hereford record number and the APHA polled numbers were shown on the American Polled Hereford Association certificate. Thus Polled Herefords became the only species of livestock in the world that was required to have two certificates issued by two associations in order to be acceptable.

Although several attempts were made to remove the burdensome yoke of double recording from the necks of Polled Hereford breeders, it continued until 1962, when a joint agreement was made between the American Hereford Association and the

American Polled Hereford Association for issuing what was known as a single joint certificate, endorsed by the secretaries of both of the associations. This single joint certificate was the first step on the long, long road toward establishing credibility for Polled Hereford breeders worldwide. A more detailed account of the struggle toward recognition will be found in later chapters.

Building Blocks for the Breed

It was fortunate indeed that nature picked the two bulls Giant and Variation on which to play her trick of mutation. Both bulls were born naturally hornless from horned parents. They were rich in the royal blood of the great Anxiety 4th, acclaimed as the Hereford breed's greatest sire.

In his book *The Hereford in America*, Don Ornduff states: "In ultimate influence upon the Hereford breed in America, no contemporary herd came close to matching the contribution of Gudgell & Simpson and no bull that of Anxiety 4th 9904. Yet, neither the herd nor the bull was regarded as a standout in the Hereford industry of the 1880's and 1890's."

Upon Anxiety's death in the summer of 1890 the *Breeders' Gazette* of August 20 of that year published this résumé:

"The famous Hereford breeding bull Anxiety 4th 9904 is dead at the age of nine years. He was bred at Stocktonbury, the home of Lord Wilton, and was got by the 'father of all the Anxietys'—the Royal champion imported Anxiety 2238, short-lived but of glorious fame in show-yard and breeding pen—and his dam was Gaylass 9905, the famous 'whiteface' matron now in the Sotham & Stickney herd. Anxiety 4th early passed into the possession of Messrs. Gudgell & Simpson, Independence, Missouri, and for years he has most worthily held the post of honor in that large and magnificent collection of Herefords. Of the few sons of the original Anxiety, all of which have achieved fame in an unwonted degree as sires, Anxiety 4th stood easily first as a getter of bulls. If he had contributed nothing to the breed but the marvelous bullock Suspense, class champion at the Kansas City Fat Stock Show of 1884, and the winner of the same honors at that exhibition and at Chicago in 1885, his reputation as a sire would have been assured, but when it is recalled that the renowned pair of show and breeding bulls Beau Monde and Beau Real owe their paternity to Anxiety 4th it will be conceded that as a sire of stock-getters he will rank among the most famous in the history of the breed."

The mysteries of genetics are still not fully understood and may never be, but there is little new in the methods of men and in the games they play. When Polled Herefords were first being talked about, most conversation by competitors was in jest or derision. Few traditional cattlemen gave them an even chance to survive

Anxiety 4th 9904, father of American Herefords.

past the first few generations before being swallowed up by the mongrelized masses of commercial cattle. After a few years, however, when they were beginning to have an impact and appeared to be here to stay, the jesting turned to ridicule and then to outright criticism. In retrospect we may observe an element of overreacting on the part of Polled Hereford breeders. Responding to some of the rumors and criticisms, Warren Gammon wrote in his *Polled Hereford Bulletin* of June 26, 1919:

"Some of the breeders of horned Hereford cattle, when they found that the Polled Hereford breeders were taking away the market for their horned Hereford bulls, started a report that Giant, the first Polled Hereford bull, weighed only about 1650 lbs., and for that reason all Polled Herefords that descended from Giant would be small, but Giant sired Polled Columbus that weighed 1900 lbs. Polled Columbus sired Gabriel that weighed 2300 lbs. Gabriel sired Gabriel 38th that weighed 2700 lbs. Now Giant's blood for the last 16 years has been mixed with Ancient Briton's blood, with Anxiety blood, with Fairfax blood, with Disturber blood, with Repeater blood, and with the blood of hundreds of other noted animals, and if his blood was so strong that he could dominate the strong blood of all of those noted animals, it does not reflect much credit on those famous animals."

Gammon's Motive

In the broad sweep of history, the success of a nation depends upon the leadership quality of her great men. In like manner, the success of a breed is based on the foundation of its great sires. To allow sires to express their greatness, they must be located and used.

In the first few generations this was a relatively easy matter because the number of traits had been narrowed to two—they must be registered Herefords and they must be naturally polled. As the breed expanded in number and the genetic base became broader, breeders were ready to add a few more traits to their selection criteria.

In the formative years from 1901 to 1922, the beef cattle world saw the dawning of a new era. Beef had become the preferred food for American dinner tables and such demand assured a bright future for cattlemen whose object was to produce quality beef. Although it was just prior to the initiation of the United States Department of Agriculture's meat grading standards, consumers were becoming quality conscious.

The introduction of the three English breeds, Herefords, Shorthorns and Angus, had improved the carcass quality of the Longhorn and given the American consumer a sampling of something he had learned to like and prefer. He was willing to pay for a better product, thus the cattleman had the added challenge of producing prime beef profitably.

With the mixture of the new breeds and their added improvement in quality and performance, the time was right to introduce the idea of increased efficiency by removing the horns that served no useful purpose.

Warren Gammon, the breed's founder, was an unusual man of varied interests and ideals. He was a deeply religious man who actively campaigned against the use of alcohol and supported the prohibition movement to prevent its manufacture and sale. He was interested in humane movements to prevent cruelty to man and beast alike. Through his monthly *Polled Hereford Bulletin* which was the first kind of breed communication, he expressed his motivation on several occasions. He believed there was no longer the need for horns as defensive weapons and that in a civilized world they were dangerous and contributed only to misery and suffering on the part of both man and animal.

The following is a *Polled Hereford Bulletin* written in August, 1919.

Is It Morally Right to Use a Horned Bull?
By Warren Gammon

"In considering this question, we should first inquire what the old cow and her progeny have done for man; second, we should consider whether or not there is any moral obligation upon the part of man to treat the old cow and her progeny mercifully and kindly.

"First, we say that more than 100,000,000 people in this country are being supplied with milk, cream, butter and cheese by the American cow. She not only supplies every American table with these necessities and luxuries, but the old cow and her progeny are feeding the nation and the world with the richest, most delicious and choicest meats for 365 days in every year. She and her progeny have supplied the nation and the world with hides that have been manufactured into leather and the leather into shoes to protect feet from the thistles, and thorns and briars in summer, and snow and frost in winter.

"Now, when we think of all the foods that have been made delicious by the use of this milk, cream and butter; when we remember that the world has been feasting upon the flesh of those animals for a thousand years, even the hair from their hides has been used in the plastering in the walls of their houses, and the bones of their bodies have been ground into bone meal, and fed to the fowls that supply our tables, even her hoofs have been used in making glue to be used in the manufacture of furniture; when we think she has manufactured the corn stalks and straw, cheap and nearly worthless feed of all kinds into the choicest kind of human feed, that she has enriched the soil of our farm, that she has gathered her own food for more than one half of the year; when we think of the great financial blessings she has bestowed on her owner and on the human family; when we think of all the mortgages she has helped to pay off, and the homes she has helped to beautify, of the children she has helped to educate, and of the numberless blessings she has bestowed upon the human family for the last thousand years; when we think of how dependent the whole human family is upon the old cow and her progeny, and that it would be next to impossible for us to live without her; when we think of all her virtues and merits, we feel like taking off our hat in her presence, and saying, 'Of all the animals on earth, she should be acknowledged as queen of them all.' The whole world should acknowledge her as a great public benefactor.

"When we consider all of her merits, we are forced to conclude that there is no species of animal on earth that is more entitled to sympathetic and kind treatment or that has greater claims on our admiration than the American cow and her progeny.

"Now the old cow and her progeny in America are owned, controlled and dominated by the American people, a people who boast of their generosity, sympathy and kindness. Now I want to ask the reader of this article if he thinks that after all that the old cow has done for humanity that she is receiving a kind, sympathetic and humane treatment at the hand of her owner? The American farmer and breeder after being raised in this land of Bibles, churches and sabbath-schools, takes down his Bible and reads from its sacred pages, 'That a merciful man is merciful to his beast.' He then calls in his neighbors and friends and he drives that grand old cow and the balance of his herd into the corral and one by one he drives into the dehorning chute, and after binding them so firmly they cannot move, he takes his saw and saws their horns and a part of their head off. This operation always causes the animal extreme pain, often causing them to faint away. No man can witness this operation and watch the blood as it streams from their heads, and listen to those heart rending shocking bawls of the animal, without feeling that the dehorning of cattle is a barbarious, heathenish practice and shocking to our feelings in the extreme, and that it is degrading to the men that practice it.

"Now we think that this is not the kind of a reward that the grand old cow and her progeny are entitled to. The whole human family has the most supreme contempt for when we dehorn we imitate the dog that bites the hand that feeds him, but when we watch the breeder as he dehorns the old cow that has conferred so many blessings upon her owner, it causes us to have great respect for a dog.

"How strange it is that farmers and breeders should practice such cruelties when it is unprofitable, unwise, unmerciful, unpleasant and unnecessary, and it is degrading to the owners and damaging to the animals, and has a contaminating influence on all that witness it.

"Now dehorning is not rare, a very large per cent of the

52,000,000 head of cattle in this country experience this horrible suffering caused by dehorning. If all of the cattle that submit to this cruel, inhuman operation each year were placed side by side in a row, and each animal was given three feet in which to stand, the number dehorned each year would make a row of cattle that would reach from the Atlantic to the Pacific Ocean. A very strange thing is why farmers will practice a thing that is unwise and unprofitable and wholly unnecessary.

"Now since the discovery of breeding the horns off cattle by sport, freak or variation, it would seem wholly unnecessary to dehorn by means of the saw, for the reason that every breeder is aware of this one fact, that there are three laws governing breeding.

"One is that like produces like, a second is the law of atavism or 'taking back,' and the third is sport, freak or variation and this law is as fixed and certain as that like produces like. Therefore, there are no cattle on the face of the earth but what can easily and safely have the horns bred off. Therefore, the last excuse for this unhuman, barbarious practice is gone."

Giant, the Foundation Sire

There were giants in those days. It was the dawn of a new century and a time for new ideas and innovative action. It was the day of the rugged individualist ready to accept any challenge. What more appropriate name could be found for a new bull calf, just a little bit different and destined to be the progenitor of a new race of cattle?

Giant was born May 3, 1899. He doubtless was not an impressive young bull for he was early consigned to the ranks of the commercial herd. Or, it could be that the absence of horns was considered a serious defect and eliminated him from consideration as a herd sire. Many breeders considered the presence of horns a masculine trait and accused polled bulls of all sorts of odd characteristics. It rarely, if ever, occurred to these men that if horns were a sign of masculinity then, in order to be consistent in theory, it would follow that horned females would be lacking in femininity and reproduction qualities.

Before proceeding to the other foundation sires and their principal breeders it will be of interest to recognize a little-known fact about the story of Giant. He was bred by O. F. Nelson of Hiawatha, Kansas, a few miles west of Kansas City, and only a stone's throw from the home of his once famous great, great grandsire, Anxiety 4th.

As a yearling bull, Giant was sold to a commercial cattleman who subsequently bred him and produced one crop of calves. Giant was himself a scurred bull, but most of his offspring were born and developed without horns. The absence of horns on his calves was obviously too revolutionary and untraditional for the man for it provoked him to return Giant to Nelson and threaten legal proceedings to recover his purchase price. It was soon after the return of Giant that Nelson received the inquiry from Gammon about his search for naturally hornless offspring of registered Hereford matings.

The progress of Polled Herefords was necessarily slow for several years, owing to the fact that most of the new blood was introduced through the mating of polled bulls with horned cows, and the percentage of polled offspring varied greatly. From time to

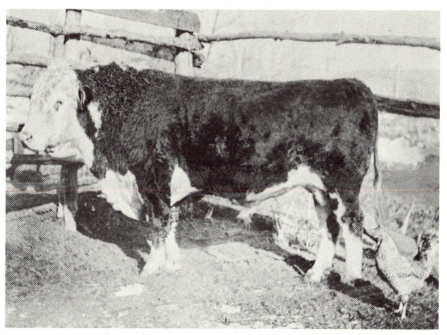

Giant, foundation sire of Polled Herefords.

time, additional purebred Polled Herefords that have appeared in herds of horned cattle have been used in developing the polled breed. While some success has crowned the efforts of breeders to secure animals possessing the polled character through the mating of polled females with horned bulls, most of the development of the polled strain has resulted from the use of polled bulls.

The first volume of the double-standard Polled Hereford herd book was issued in 1913. It contained the names of 235 breeders and the record of 2,250 cattle. Of the 235 breeders, 62 were located in Iowa, which was the leading Polled Hereford breeding state for many years. Since 1956 the leadership in registrations has rotated between Missouri and Texas.

Giant was used by the Gammons for several years and then sold to G. E. Ricker of Ashland, Nebraska. He was used mostly on horned cows. Giant sired nine sons that became well-known in the breed and were considered to have made a substantial contribution toward securing the breed's future. The best known of his sons and the pioneer breeders that used them follows.

Noted Sons of Giant

Polled King 159709 (12), Renner Stock Farm, Hartford City, Ind.; Polled Assurance 193115 (29), E. W. Gammon, St. Charles, Ia.; Polled Admiral 2nd 230299 (43), J. W. Wyant, Blythedale, Mo.; Polled Peach 223239 (49), J. W. Wyant; Polled Columbus 228226 (54), D. Walton, Denver, Ill.; Polled Modern Briton 256951 (67), D. C. Maytag, Laurel, Ia.; Royal Giant 255720 (77), H. R. Ackland, New Virginia, Ia.; Polled Phillippi 249335 (53), Bent Hereford Livestock Co., Hazard, Nev.; and Polled Carlos 249334 (118), H. L. Leonard & Son, Waukee, Ia.

Variation

Another naturally polled bull, second only to Giant in his contribution to Polled Herefords, was Variation 152699 (14),

This stone marker and plaque on the original Gammon farm near Des Moines, Iowa, commemorates the first planned mating of Polled Herefords.

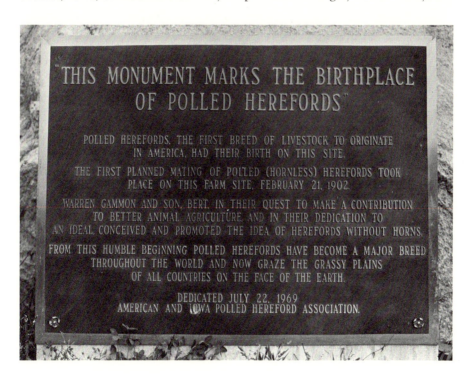

"THIS MONUMENT MARKS THE BIRTHPLACE OF POLLED HEREFORDS"

POLLED HEREFORDS, THE FIRST BREED OF LIVESTOCK TO ORIGINATE IN AMERICA, HAD THEIR BIRTH ON THIS SITE.

THE FIRST PLANNED MATING OF POLLED (HORNLESS) HEREFORDS TOOK PLACE ON THIS FARM SITE, FEBRUARY 21, 1902.

WARREN GAMMON AND SON, BERT, IN THEIR QUEST TO MAKE A CONTRIBUTION TO BETTER ANIMAL AGRICULTURE, AND IN THEIR DEDICATION TO AN IDEAL, CONCEIVED AND PROMOTED THE IDEA OF HEREFORDS WITHOUT HORNS.

FROM THIS HUMBLE BEGINNING POLLED HEREFORDS HAVE BECOME A MAJOR BREED THROUGHOUT THE WORLD AND NOW GRAZE THE GRASSY PLAINS OF ALL COUNTRIES ON THE FACE OF THE EARTH.

DEDICATED JULY 22, 1969
AMERICAN AND IOWA POLLED HEREFORD ASSOCIATION.

Excellent Ion, sired by Variation.

Bullion 4, record selling bull, 1919—$9,500.

dropped from horned sire and dam in the herd of John G. Thomas, Harris, Mo. Variation was secured by the Mossom Boyd Co., Bobcaygeon, Ont., one of the most prominent among the early breeders of Polled Herefords. Bullion 4th 428447 (3062), one of the foremost Polled Hereford sires, bred by the Mossom Boyd Co., was descended, through both sire and dam, from Variation. Giant and Variation were the two most influential sires founding purebred Polled Herefords.

Mossom Boyd

Mossom Boyd was one of the first breeders of prominence to engage extensively in the production of Polled Herefords. He was a close student of animal breeding, and entered enthusiastically into the work of developing and improving the hornless white-faces. One of his first herd sires was Wilson 126523 (7), a naturally polled bull from horned sire and dam. His sire was Bismarck 74564 that traced to Sire Richard 2nd and Lord Wilton. His dam was similarly bred. Variation was, however, the bull that gave the herd most of its prestige. Two of his most noted sons, used in the Boyd herd, were Variation 3d and Excellent Ion. Bullion 4th, sired by Formation 2d, a son of Excellent Ion, in the hands of the Renner Stock Farm, Hartford City, Ind., became the most noted of Polled Hereford bulls.

Renner Stock Farm

The Renner Stock Farm herd, established by Benjamin Johnson at Hartford City, Ind., was another pioneer Polled Hereford herd and has continued to hold a leading place in the industry for many years. One of the early herd sires was Polled King by Giant. His sons King Edward (78) and Polled King Wilton (87), and Polled King Giant by King Edward were used for several years. They were followed by Polled Ito by Polled Admiral 2d, Dominion by Variation 3d, and Bullion 4th. Repeater 96th by the grand champion Repeater 7th, a horned bull, also was in service.

A dispersion sale was held in 1919 at which 73 head averaged $931, an unusual average for that day. The eight bulls averaged $2,109. Bullion 4th went to W. A. Wilkey & Co., Sullivan, Ind., at $9,500. Repeater 96th went back to O. Harris & Sons at $2,500, and his son Ion Repeater, a polled bull, went to E. W. Vanderwater, Orient, Ia., at $1,000. Fred O. Hagemann, Mt. Vernon, Ind., paid $1,500 for Bullion 24th by Bullion 4th. The top female was Pretty Pansy by Dominion with heifer calf by Repeater 96th. She went to F. A. Stimson, Huntingsburg, Ind., at $2,000. The 65 females averaged $786. The Renner Stock Farm was purchased by F. A. Stimson, who secured a half interest in Bullion 4th from Wilkey. A joint show herd was maintained for many years. Bullion Garfield by Bullion 4th was grand champion at the National Polled Hereford Show at Des Moines, Ia., in 1923 and 1925, and winner in many other shows in competition with horned Herefords. Foundation 25th by Foundation 4th, bred by the Mossom Boyd Co., and owned by the Renner Stock Farm, was grand champion at Des Moines in 1924.

J. E. Green

J. E. Green, Muncie, Ind., was one of the most enthusiastic pioneer advocates of Polled Herefords. Much of his foundation stock came from the Mossom Boyd and Renner herds. Polled King LeRoy and his son King Jewel, and Polled Peach 10th by Polled Admiral 2d were among the stock bulls used by Green.

Jenkins Bros., Orleans, Ind., maintained a small herd since Polled Herefords began to attract attention. Their first herd bull was a son of Prince of Iowa.

In later years they used Polled Euchre by Polled Pride, and Euchre Lad by Polled Euchre. At the 1925 Des Moines show they purchased Perfection Gem by Gemmation 2nd from G. E. Pettigrew & Son, Flandreau, S. D.

F. P. Bieth

F. P. Bieth, Joliet, Ill., was a prominent figure among the early Polled Hereford breeders and attracted much attention to the polls through his excellent exhibits at the International. His early herd bulls were Mutation by Variation; Adastion, a grandson of Wilson, and King Jewel 3rd by King Jewel.

R. T. Painter

R. T. Painter, Stronghurst, Ill., another pioneer Henderson County breeder had one of the leading herds of Polled Herefords. He started with Polled Victor by Polled Success 4th and followed him with three of his sons—Polled Dandy, Victor's Dandy and Polled Count. Then he used Polled Anxiety by Polled Climax and Polled Duke by Polled Count. In 1916 Painter paid Glaves & Painter, Lewistown, Mo., $1,005 for the two-year-old heifer Polled Marvel by Polled Pride and mated her with Polled Duke. The resultant calf was Marvel's Pride 2d that became the chief herd sire for several years.

Painter and R. C. Glaves operated a partnership herd on the farm of the latter at Lewistown, Mo., for several years, using Polled Climax (951), Polled Victor, Admiration by Polled Pride, and Polled William by Polled Ito. The herd later was owned by Glaves, who continued to use Polled William.

George T. Rew

George T. Rew, Silver City, Ia., was a prominent breeder of horned Herefords before he changed to the use of polled bulls. Back in the early 1890's he used Ned Brandon whose sire Lord Brandon was a double grandson of Anxiety 4th 9904; and Lord Mulberry by Earl of Shadeland 30th. On the foundation laid by these bulls he used Polled Victor, Excel Variation by Variation 3d; Polled Atlas G, a grandson of Giant; Polled Ham, a grandson of Polled Admiral 2d; Wizard Archer by Jolly; Pharoah Boy, a

Pearl, record selling female, 1919—$3,900.

King Jewel 5th, first Polled Hereford reported to have been sold for $1,000.

grandson of Dominion; Polled Richard by Polled Climax, Polled Gradest by Polled Duke, and Prime Defender by King Jewel 4th.

Star Grove Stock Farm

The fame of the Star Grove Stock Farm herd at West Liberty, Ia., owned by P. M. Schooley & Sons, rested largely upon Star Grove, a son of Echo Grove, chief herd sire for a number of years. Bullion 14th, a son of Bullion 4th, headed the herd for a period.

G. E. Pettigrew & Son

One of the early leading herds of Polled Herefords in the northwest was that of G. E. Pettigrew & Son, Flandreau, S. D. They used Polled Anxiety by Polled King 3d, but the reputation of the herd is due largely to the service of Canadian and Gemmation 2d. Their show herd was one of the most successful at the National Polled Hereford Show at Des Moines, Ia.

Polled Herefords in Nebraska

E. H. Gifford, Lewiston, was among the first to promote the Polled Hereford cause in Nebraska, and his herd remained in prominence for many years.

The principal bulls used in the Gifford herd were Polled Echo by Polled Admiral 2d, Polled Pride by Polled Victor, Polled Richard, Polled Euchre and Model Mischief. Polled Euchre went to the herd of Jenkins Bros., Orleans, Ind. Model Mischief is out of a dam by Mousel's Beau Mischief.

Polled Herefords in Kansas

Wallace Libbey, who had been a pioneer breeder of horned Herefords at Ottawa, Ill., moved to Larned, Kans., in 1909 and began using Polled Hereford bulls. In a few years Larned became the center of one of the leading Polled Hereford breeding communities in the country. Mr. Libbey used Polled Plato (884) whose blood remained strong in Kansas herds. Later he moved to Maxwell, N. M., where he continued to breed Polled Herefords for a number of years.

John M. Lewis, Larned, Kans., was one of the prominent breeders of Pawnee County since 1910. His herd was one of the largest in the country. Polled Plato 8th and Polled Plato 18th by Polled Plato, and Polled Eclipse by Polled Idlewilde D by Polled King 3rd and Pawnee Chief by Excellent Ion have been his leading herd bulls. Polled Perfection, a grandson of Polled Plato, and Model Perfection by Polled Perfection were used for several years. The Lewis herd is one of only a few pioneer herds still in existence and is operated by sons Walter and Joe as a partnership.

Polled Herefords in Missouri

One of the first herds of purebred Polled Herefords in Missouri was that of J. R. Hill, Norborne, Mo. Hill used King Giant by King Edward 7th, King Ruler by King Jewel and his sons, and Beau by King Giant.

Polleds in the Range Country

In the range country, Polled Herefords kept pace with their progress in the cornbelt states. Prominent early breeders were W. N. Shellenbarger, Oklahoma City, Okla.; M. W. Hovenkamp, Keller, Tex.; and William E. Wallace, Midland, Tex., establishing breeding herds while Polled Herefords were still few in number, and through these herds Polled Hereford bulls found their way into many range herds. Wallace maintained one of the largest herds of Polleds in the southwest for a number of years. In 1915 he purchased Polled Victor (1929) and his son Polled Victor B from W. H. Campbell, Grand River, Ia., for $2,150. Sons and grandsons of these two bulls were in service for many years.

Among the later herds of Polled Herefords in Texas that

achieved prominence were those of Henry & McCloskey, San Antonio; Burleson & Johns, Whitney; Keith Hereford Farm, Wichita Falls, and P. S. Kendrick, Albany. Mr. Kendrick was elected president of the American Polled Hereford Breeders' Association in 1925.

Polled Herefords in the West

A number of herds of Polled Herefords gained a foothold in the western states. One of the most prominent was that of John B. Bowman, McIntosh. Babbitt & Cowden, Phoenix, Ariz., had one of the leading herds of the west. Many Polled Herefords went from the herds of the middlewest to California in the early years, establishing a number of herds there. The Roosevelt Live Stock Co., Cleveland, Ida., A. C. Bayers, Lavina, Mont., W. S. Haley, Terry, Mont., and Taylor Ranch Co., Stanford, Mont.

Polled Hereford Breeders in the East

Hugh A. Coyner, Waynesboro, Va., had a herd of Polled Herefords for several years. It was headed by Polled Alfred that later went to head the herd of J. F. Patterson & Son, Bedford, Va., later to become one of the leading herds of the state.

T. W. Herron & Son, Chandlersville, were among the oldest breeders of Polled Herefords in Ohio. Their principal herd bull was Prince Albert by Bullion 4th.

One of the pioneer herds of Polled Herefords in the southeastern states was that of John F. Cason, Murfreesboro, Tenn., who secured his foundation stock, including his first herd bull, Polled Beau Real by Polled Admiral 2d, from J. W. Wyant, Blythedale, Mo.

M. H. White, Olive Branch, was the leading pioneer breeder of Polled Herefords in Mississippi. His herd was headed by Bullion 26th by Bullion 4th and Allen Bullion by Bullion 26th.

The Beginning of Polled Herefords in England

Uprooting the past history of Polled Herefords reveals true stories of drama and intrigue unparalleled in fiction. The ingenuity and resourcefulness of stockmen spurred by the desire to profit combined to create motivations beyond restraint.

Such is the case for the first introduction of Polled Herefords back to Old England, the homeland of Herefords.

The Hereford breeders of England had long been skeptical about the origin of the polled gene in the breed and had not recognized Polled Herefords in the Herd Book from the breed's beginning in America.

What in effect was ruling out the legitimacy of Polled Herefords created a ripe situation for some ingenious person to beat the rule. According to a report from the English Hereford Herd Book Society Colonel David Talbot-Rice imported some semen in 1948 from a bull CMR Advance Domino 81st and by this bull he produced a bull calf out of an English Hereford cow and named him Coln Arthur.

Although Coln Arthur never attained legitimacy nor has any of his progeny ever been admitted to the pure stud book of the

Hereford Herd Book Society, he paved the way for later importations and today approximately one-half of all Herefords recorded in England are polled.

The story of Coln Arthur is a rare one indeed, matching—if not surpassing—that of Giant, the foundation sire of the breed.

The late Sir John Hammond of Cambridge University was interested in experimenting with the Inter-Continental exchange of frozen semen, a technique which was then in its infancy. He succeeded in acquiring some frozen semen from the bull CMR Advance Domino 81st in the U.S.A. It is believed that this semen was smuggled into England, contrary to the existing veterinary regulations, by an RAF pilot in a Spitfire fitted with long-range fuel tanks to cross the Atlantic.

This was the semen used by Colonel Talbot-Rice on 30 purebred, but unregistered, Hereford cows and although all 30 cows were inseminated only one calf resulted, he being the polled bull, Coln Arthur.

Because artificial insemination was not then recognized and because, in any case, Colonel Talbot-Rice's cows were not properly registered, the Coln Arthur Polled Herefords have never been eligible for the Herd Book itself.

At the same time, in 1951, Mr. Cecil Evans of Wroxall, Warwick, and his brother, Leslie, used this bull on some horned Hereford cows. In 1951 Major North, a noted Hereford breeder, purchased one of Coln Arthur's sons and used him on his herd of horned Herefords.

At this time Colonel Talbot-Rice and Major North ran the Coln Arthur register themselves as they were not recognized by the Hereford Herd Book Society, who in spite of repeated requests from them refused to recommend the import of Polled Herefords.

In 1954 the Chief Livestock Officer from the Ministry of Agriculture inspected the Polled Herefords which had been bred by Colonel Talbot-Rice and Major North. After seeing both herds he expressed the view that they were of a suitable standard and that he would, therefore, recommend that they be given permission to import live animals.

They received permission to import from the Ministry of Agriculture, but were confined on veterinary grounds to importing cattle only from New Zealand. In the autumn of 1954 Major North went to New Zealand and, as he had sold his farm in England, was forced to sell the Coln Arthur cattle which he had bred. These were divided equally between Colonel Robert Henriques and Mr. Oscar Colburn to form the foundation of the Winson and Crickley herds.

Major North went to New Zealand with an order for polled cattle to be sent to Mr. Douglas McDougall of Cooper, McDougall and Robertson in England. He bought one bull and five heifers which were shipped to England. These cattle arrived at Tilbury in 1955 and were immediately recognized by the Hereford Herd Book Society as pedigree Hereford cattle, a status which had been denied to the Coln Arthur Polled Herefords bred earlier.

It was as a result of this successful importation from New Zealand that permission was then given by the Ministry of Agriculture for a "once and for all" importation from the USA and 22 cattle were imported in 1956. This second importation included cattle for a number of breeders including Mr. H. A. D. Cherry-Downes, Mr. Oscar Colburn and others.

The original 1955 importation from New Zealand of the bull Toko Excelsior and five heifers constituted the first entries in what was then Volume B of the Herd Book. Before that time the Coln Arthur register had been kept by Colonel Talbot-Rice and Major North, who between them had personally conducted all the correspondence with the Ministry and the Hereford Herd Book Society. In 1954 Mr. Douglas McDougall offered the services of his office and Mr. Dolan to keep the records. Between 1951 and 1954 Messrs. Talbot-Rice and North held many meetings of which there is no record except perhaps at the Ministry of Agriculture.

The Polled Hereford breeders of Great Britain Limited was founded in 1955 by the following: David Talbot-Rice, Douglas Sidney Arundel McDougall, Richard Jopson, Cecil Henry Evans, Oscar Henry Colburn, Leslie Edwin Evans, and Robert David Quixano Henriques.

In March 1957, an importation was made by Oscar Colburn, of Crickley Barrow, Cheltenham. He selected BPF Pawnee Perfect from the Bushy Park herd of John Royer, Jr., Woodbine, Maryland. BPF Pawnee Perfect, a Polled Hereford bull, became one of the most successful breeding bulls in England Hereford history.

H. A. D. Cherry-Downes, of North Clifton Manor, Newark Notts, imported the bull, Gay Hills Victor 46 in March 1957, that further enhanced the progress of Polled Herefords in England.

John Young, of King's Lynn, imported the popular Gay Hills Victor 78, that ultimately became one of the great foundation sires of Polled Herefords in England.

4. Scanning the History

National Events
Annual Meeting 1922
The First Great National Show

National Events

The spirit of the breeders of yesteryear and the nostalgia of days gone by is captured in the written accounts of the national gatherings that became so important to Polled Hereford breeders.

In reading the stories written about past national sales and shows, one is impressed with the emphasis placed on them by the early breeders. These events seemed to provide the opportunity for fellowship and comradeship so needed in a world where transportation and communication were not as yet sophisticated. The annual gatherings served as a source of encouragement and reassurance to a fledgling group struggling to overcome inhibitions and complexes. Each breeder could return to the loneliness of his ranch and attack his daily routine, buoyed by the exchange of ideas and the realization that others were sharing his dreams and aspirations of making Polled Herefords a great breed. He also became acutely aware that others were confronted with the same disappointments and momentary misgivings that he had.

All members of this small, innovative group must have been determined to win over the formidable odds of prejudice and the great resources of other breeds. They had their times of despair, but the recorded accounts of the national gatherings are marked with a spirit of optimism in spite of the circumstances.

Many clichés have been written about the secrets of success
and the key element of determination. Failure is never final; it
can be only an interlude to success. One never loses unless one
quits too soon. Such was the dauntless spirit of the early breeders
of Polled Herefords.

It would be a cumbersome task to report in detail accounts of all
of the shows and early gatherings of Polled Hereford breeders. In
order, however, to give the student of Polled Hereford history an
understanding and a feeling for the people and the problems of
those early days, brief accounts of major occurrences in 1922 and
1923 appear in subsequent pages and are of particular importance.

The earliest reports of Polled Hereford events were carried in
the *American Hereford Journal* which was the first publication to
represent the breed. The *Journal* was privately owned, having
been founded by its publisher Hayes Walker, Sr., May 1, 1910.
This single publication did an adequate job of publishing accounts of the happenings of both the horned and polled segments
of the breed. It was purchased by the American Hereford
Association in March, 1961.

Frank Farley, Sr., longtime field representative for the *American Hereford Journal* and co-founder of the *Polled Hereford
World* magazine in 1947, covered most Polled Hereford events.
The reports and stories of the Polled Hereford National Shows and
annual meetings were probably from the pen of Frank Farley, Sr.
Farley spent more than 50 years as a field representative, traveling
extensively covering Hereford activities and writing about the
breed. He was posthumously inducted into the American Polled
Hereford Association's Hall of Merit in 1973 for his dedication
and contribution to the breed. The late Hayes Walker, Sr., was the
recipient of the same honor in November, 1975.

Special plaque presented posthumously to Hayes Walker, Sr. Inscription reads: The Polled Hereford breeders of America dedicate this memorial tablet in grateful recognition of his thirty-four years of service to our industry. As our valiant champion whether we fought the tides of prejudice or rode the crest of popularity, always our constructive critic when we needed guidance. Our generous supporter in days of financial stress, our wise counsellor when we sought advice, and, withal, a warm-hearted personal friend to each of us. The memory of this helpful advocate will continue to be cherished by our fraternity through the years. Unanimously voted at the annual meeting of the American Polled Hereford Breeders Association at Atlanta, Georgia, October 19, 1944. Shown seated are, left to right, B. O. Gammon, Ia; A. G. Rolfe, Md; John Lewis, Kans.; Mans Hoggett, Tex.; Standing, left to right, are: Don Chittenden, Ia.; Hubert Mullendore, Ind.; John Rice, Wyo.; Jim Gill, Tex.; M. P. Moore, Miss.; Clifton Rodes, Ky.; Hayes Walker, Jr., Mo.; Dinty Moore, Ga.; John Trenfield, Tex.

1951 American Polled Hereford Association Board of Directors.

ALF Battle Mixer 30, 1953 National Polled Hereford Champion Bull.

ALF Monarch 37, 1956 National Polled Hereford Champion Bull.

CEK Mixer Return, 1958 National Polled Hereford Champion Bull.

Golden Diamond, 1961 National Polled Hereford Champion Bull.

W. Lamplighter E 43, 1965 National Polled Hereford Champion Bull.

Walter Lewis, Larned, Kans., presenting the Australian John E. Rice Memorial Trophy to R. E. "Pat" Connolly, St. Helena, Calif., for breeding the December, 1969, National Champion Beau Mischief 3.

Lineup of yearling bulls, National Polled Hereford Show, Atlanta, Ga., December, 1969.

Passing of the gavel from National APHA President, Lynn Storm, Dripping Springs, Tex., to new President, Leon Falk, Jr., Pittsburgh, Pa., December, 1969

A national winning get-of-sire, shown by Alfalfa Lawn Farm, Larned, Kans., at National Polled Hereford Show, Atlanta, Ga., December, 1969.

National Polled Hereford Sale, December, 1969, Atlanta, Ga.

December, 1969, National Champion Bull, CPH Beau Mischief 3, bred by Connolly Polled Herefords, St. Helena, Calif., and shown by Falklands Farm, Schellsburg, Pa.

Mrs. R. W. Jones, Jr., Leslie, Ga., receiving the coveted Hall of Fame award posthumously for her late husband, a pioneer in performance testing, at the National Awards Banquet, Denver, Colo., January, 1971, presented by Otha H. Grimes, Tulsa, Okla.

Annual Meeting 1922

It was "England's finest hour," in the glowing language of Winston Churchill, when she was bruised and bleeding from the Nazi bombers of Germany but refused to give up. Only on a few occasions are such legendary happenings dramatized with such eloquence. If any group of people was to be singled out for similar recognition it should be the cattlemen of the world. It is unlikely that any other fraternal group is subjected to the hazards of economic gyrations, fluctuations in supply and demand, pestilence, drouth, flood and disease. Not once but several times in the life of the average career cattleman, he may have to, as Rudyard Kipling put it, "Watch the things you gave your life to, broken, and stoop and build 'em up with worn-out tools."

There was great optimism among cattlemen following the end of World War I in 1918. Unusually good prices were being paid for all agricultural products, especially beef. Purebred breeders were enjoying record prices for their seed-stock. The air was adrift with optimism. At the National Polled Hereford Sale in 1918, the overall average paid was $1,253 per head, the highest on record.

In 1919 the average was $1,144. The 1920 National Sale averaged $1,020, but by 1921 the average had dropped to $346 per head. The greatest shock, however, was in 1922 when the average price dropped to $221 per head. This represented a drop of more than 82 percent in four years.

In spite of the looming economic storm clouds and the backdrop of gloom and doom, Secretary Gammon reported Polled Hereford Week in 1922 to be the most successful ever staged. "Why?," reported Gammon, "Because there was evidenced, from the first hour of the occasion until the last visitor departed, a spirit of breed confidence which, in view of what breeders had endured in recent months, was nothing short of marvelous. Not a single 'croaker,' not a man who talked 'quit,' not a single note of pessimism, not a man but was facing forward and confidently expecting better days in the near future. When men can come through such days as mark the recent past with this spirit of optimism and confidence, with a determination to stand by their guns, with such evidence of a willingness to cooperate for the good of all, then the project with which they are connected is in a stable and safe condition. Polled Hereford Week 1922 proves that the breed is destined to a remarkable expansion during the next season. It was the test of fire and the steel proved its temper."

The First Great National Show

Although there had been twenty-one years of breed history written before the year 1923, that year was one of great significance to the breed of Polled Herefords. It was a well-known fact that the dedication and determination of Warren Gammon had great influence on the surge of the breed during his lifetime. Through his dynamic personality and prolific pen he kept the breed before the public's eye and was a great morale builder for all the breeders when times were difficult. Such were the times from 1919 to 1923.

It was a great setback to the breed when he died in 1923. It was also a test of the leadership of young Bert Gammon and the breed's other leaders. The year 1923 did not find them wavering or weakening. It was only the beginning of what has grown into the greatest cattle show and promotional event for a single breed in the annals of beef cattle.

The National Polled Hereford Show had its birth about the same time of the breed founder's death. It has now become the oldest and largest traveling breed show in history.

An obvious attribute to even the casual observer is the

optimistic and positive attitude of Polled Hereford breeders. The pens of the writers covering the National Polled Hereford events capture and record this characteristic of the Polled Hereford industry, one which has been carried down to this day. Just as the polled trait is dominant in the breed, so is optimism and the ability to overcome prejudice, adversity, and disappointment inherent in Polled Hereford breeders—a trait that renders any defeat as only temporary.

This priceless quality in man, which is essential to success, may be credited to the breed's leaders. Men such as Warren Gammon, the father of the breed and son, Bert, long-time secretary; Hayes Walker, Sr., founder of the *American Hereford Journal*, and Frank Farley, Sr., the long-time field representative of the *American Hereford Journal*. They contributed so much, for so long, to the advancement of both segments of the Hereford breed.

The Sale Ring at first National Show.

First place females, 1942 National Show, Atlanta, Georgia.

1954 National Polled Hereford Show and Sale. Left to right: John Shiflet, Red Rock, Oklahoma; Dick Hibberd, Imbler, Oregon; F. L. Robinson, Kearney, Nebraska; Joe Largent, Newkirk, Oklahoma.

1954 National Polled Hereford Show and Sale. Left to right: D. O. Geier, Banner, Wyoming; Jim Gill, Coleman, Texas; M. P. Moore, Senatobia, Mississippi.

5. Brush Fires and Pot Shots

Polleds Denounced

Polleds Denounced

Never in the history of any breed has there been expressed more determination by the advocates or more prejudice, emotion, and reaction on the part of the adversaries than in the beginnings of Polled Herefords. Early Polled Hereford breeders ignored the lowly birthright foisted upon them by traditional Hereford breeders. They went about with head high but an inward feeling of inferiority, which expressed itself at times with open aggressiveness and rash predictions. The horned breeders maintained a proud stance and an air of ridicule and held onto the horned concept with an almost religious tenacity. There was little question that very early battle lines were drawn, and there were only a few who recognized any middle ground.

Many breeders attempted to breed both horned and polled cattle, maintaining a herd of each, but all would eventually abandon either the polled or the horned cattle and concentrate on one or the other. No doubt outside influence from other breeders or pressure from one of the associations contributed to the incompatibility of maintaining herds of both horned and Polled Herefords. In 1929 a series of letters appeared in the *American Hereford Journal* that seemed to typify the attitudes of breeders of the two segments toward each other. W. J. "Willie Joe" Largent was one of the colorful show men of all times. A keen observer, a master breeder, and an ardent promoter of the horned segment, in later years he became a Polled Hereford breeder and a nationally renowned judge. After his conversion Willie Joe is said by his friends "to have fought as hard for Polled Hereford prominence as he did against them" in his earlier years.

The following exchange of letters presents to the reader a vivid picture of the attitudes between the segments during that time:

Largent Denounces the Polls

To The *American Hereford Journal:*

Received the Feb. 15 issue of the *American Hereford Journal,* The Polled Hereford Number, today, and as I have been inside all day nursing a sprained hip, have been amusing myself at the many assertions made by breeders of Polled Herefords and how few facts occurred in their articles. All in the world they have to say against the horned Hereford is the fact that he has horns and all they can say good about the Polled Hereford is that he does not have horns. Every article is full of the same old story of shrinkage and loss of blood and the awfulness of dehorning.

As I have written through your columns before, dehorning is one of the simplest things any cowman ever had to do. All that is necessary to dehorn a calf, listen, Mr. Polled Breeder, once and for all, is to apply a stick of caustic to the button when the calf is around one month old, or take a pocket knife, lift the button and sear with hot iron. Two men can handle 300 head of calves a day at a cost of $4. "Terrible Expense," and when you do this you are not sacrificing all this good beef conformation that we horned breeders have been all these years perfecting.

We quit dehorning years ago because every once in a while some fellow that did not know much about beef conformation would ask us if the calves we had were Polled Herefords and we got afraid some of our many customers might think we were breeding them and our good trade would of course stop. I have seen very few polled cattle that were good cattle and I have seen a good many and have made many of the best shows of the country and have seen the best that the breed produces, but one thing I have noticed that when you do get one of those freaks that does look fairly good he does not breed true to his own conformation.

If the Polled breeders have another good point other than that they breed the horns off them (I do not even consider this a good point) it is that they have increased to I believe 65,000, according to their records. Well I just did not know that we had that many cowmen that don't know good cattle when they see them or else are not trying to breed the good kind.

I have heard for the past several years that all the Polled breeders were going to get together and from these 65,000 head of polled cattle pick out a show herd, put it under one name and go out for bear. I was in hopes that idea would materialize at this last Des Moines show and that they would send them on down to Fort Worth, Texas, to clean up on us horned Hereford men.

We have a breeding herd of 65 cows. I would like to select a show herd for next year from this small herd to show against what they would select from these 65,000 head. Oh, you say, there is so much in fitting. Well, you have to have something to fit when you start if you expect to get anywhere. We feeders find cattle that we cannot fit and that is the trouble all polled breeders find.

I notice in the *American Hereford Journal* a comment that the cattle did not carry the bloom at the Des Moines show. That is quite singular, when the country from where those cattle came will grow more of the kind of feed that it takes to fatten show cattle on one acre than we can grow here on fifty. Is it possible that there is not a man breeding Polled Herefords that does not know how to mix a simple ration of corn, oats and bran?

Have also heard the story about Polled Herefords not getting their dues in the show ring against horned Herefords. I do not think this a just claim because I know that many of our best judges are honest men, and then besides look at the good college judges we have that do not have any fish to fry, but always put cattle just as they see them.

As I see this Polled Hereford herd that is going out to clean the plate is not going to materialize, I just want to ask you Texas Polled Hereford breeders to come out to the different shows in Texas this fall and show your wares. If you have the best cattle then I will take off my hat to you, if you haven't then do not go back home and bring up this old story about the loss of blood and shrinkage of dehorning cattle. There is plenty of room for all of us in this old United States of America to breed the different breeds of cattle that we prefer, but when you just keep on making assertions and then will not meet us and prove that you have what you claim to have then I thought it time to just tell you what you do have.

You can fool some of the people some of the time but you can not fool all of the people all the time. One article tells of how much more a polled bull was worth than one with horns, citing again the story about all this loss of blood and shrinkage. It is just like I have said before, I had rather sacrifice a little shrinkage (but if handled properly there is not any) than to sacrifice three-fourths of the beef characteristics.—W. J. Largent, Merkel, Tex. [*American Hereford Journal*, March 15, 1929].

Sympathy for W. J. Largent

Such a painful hip that could cause so bad a slip.
Such a sad mistake did our Texas Largent make.
Let us hope great good will be in his meed,
Now that he has discovered a Polled Hereford breed.

To Mr. W. J. Largent, Merkel, Tex.:

In your weakened condition, please allow me to extend to you my heartiest sympathy. You certainly deserve the sympathy of every Hereford breeder, both horned and polled. It is quite evident from your article in the recent good Texas Edition of the *Hereford Journal* that your "sprained hip" has put you in a very serious condition.

Be assured, Mr. Largent, that my sympathy is genuine. If there is anything that will help in getting you back to your normal condition that is within my cable-tow, you have but to name it and it is yours. Yours, most sympathetically.—John C. Riggs, Dos Cabezos, Ariz. [*American Hereford Journal*, April 1, 1929].

President H. J. Smith of Polled Association
Replies to W. J. Largent

To The *Hereford Journal:*

The letter of W. J. Largent, of Merkel, Tex., addressed to you and published by you in your issue of March 15 at page 112, is before me. First of all, I want to compliment Mr. Largent upon his upstanding defense of the horned Herefords. It is such breeders, in all lines, who ultimately present to the cattle breeders' organizations of all kinds their faith in this, that or another stock. There is just one suggestion, however, that offers itself to the mind of the reader and that is that notice be taken that not only the breed defended in the article has points of merit; that, more than likely, all breeders of established lines have substantial points of merit to present in introducing or advocating their respective lines.

Breeders, in all lines of established repute, might well bear in mind the cardinal thought and, to my mind, the cardinal fact that all established lines do have points of merit and, if such is admitted, to respect the effort to promote and broadcast these points; to inculcate these in the minds of potential breeders so that the inquiring mind may hear more of merit and substance concerning any related lines. A charitable respect for the pioneers

or leaders in the advocacy of any or all lines needs encouragement.

Those of us belonging to the Polled Hereford family may point
with some degree of comfort to the fact that even the vigorous
cattle family of the Lone Star State recognizes that there is much
virtue in the Polled Herefords. To evidence this fact we may point
to the recent sale at Fort Worth at which Polled Herefords
averaged $408. This tribute was paid, in substantial form, by the
neighboring cowmen of Mr. Largent.

Obviously, these buyers, the neighbors of Mr. Largent, were not
influenced by the single point that dehorning may or may not
result in injuries to the stock, for, in all probability, Mr. Largent's
harmless method and inexpensive means of dehorning is known
to all the cattle fraternity in Texas.

We believe that we have demonstrated the virtue of the Polled
Hereford line. In 25 years we have risen from infancy to a
substantial industry, and we are showing our stock against lines
that have been developed in the horned Hereford industry for
generations. Our measurable success must be due, in a larger
degree, to the merit of our lines than to technical claims. If some
judges have favored the horned lines, others, of equal experience
and integrity, are partial to the Polled Herefords; and we are
willing to concede, generally, the good faith of both groups.
Opinions of men, in contrasting two meritorious lines, are bound
to be apart, and when honestly entertained, little is to be gained
by contending against them in the columns of the press.

From our point of view, based upon and tried in the acid test of
experience, the Polled Herefords are entitled to and have been
accorded a place of worth and distinction. This position may be
pointed to with no little degree of pride and satisfaction, and we
do not apologize for our faith in it. And, if, perchance, the horned
Hereford breeder has a like conviction, who would deny his right
thereto.

May we not, therefore, continue to live and defend the faith that we have in this industry of ours which, as already pointed out, has earned its spurs and proudly wears them. And the right we claim we are willing to accord to Mr. Largent.

And, finally, the good purposes that the horned Hereford family and the Polled Hereford family may serve are far removed from petty quarrels and differences; the aims which are to be attained is the service to the cattle industry generally. Let us all join in contributing the best of what each may give. Each by contributing to his own cause will, incidentally, advance the cause of the other. Why impair this common interest?—Heller J. Smith, President, American Polled Hereford Breeders' Association, Bellwood, Neb. [*American Hereford Journal*, April 1, 1929].

W. J. Largent in Rebuttal

To The *Hereford Journal*:

Have received today the April 1 *Hereford Journal* and have read with interest the different answers to my argument against Polled Herefords. Some of them cite their experience in the cattle business. The firm of C. M. Largent & Sons has been breeding registered Herefords 28 years and has been exhibiting for 24 consecutive years, and while I am yet a young man, I have spent 20 years in breeding, fitting and exhibiting Hereford cattle, and have had considerable experience with grade cattle and have fed carlots of cattle for both market and show.

I want to say that if all the articles that have appeared in favor of Polled Herefords in the *Hereford Journal* had been written by such men as John M. Lewis of Kansas and H. J. Smith of Nebraska, I would never had had to answer the different articles, because they are in a measure reasonable in their statements, but it was the articles written by those who made only assertions that compelled me to write. I do not believe I will get many answers to the questions I asked and to the statements I made about showing their wares; and I am, in a way, sorry that I had to show them up, as the saying is, but perhaps a lesson now and then will help. I want to answer one thing in Mr. Smith's article and that is relative to neighbors of mine here in Texas buying Polled Herefords at the Fort Worth show at the price they did. I have been in the cow business long enough to know that there are quite a few cowmen that do not know good cattle when they see them and I do not think Texas an exception to that rule, therefore do not hold me responsible for some of that buying in the Polled Hereford sale at Fort Worth. I am glad that Mr. Lewis appreciated the statement I made in my article in the March 15 issue of the *Hereford Journal* where I stated that there was plenty of room for all of us in the United States to breed the kind we prefer, and I want to say that it does not make any difference what beef breed he has if he will let them stand on their own merits, but I can't say much of the fellow that claims what he does not have.—W. J. Largent [*American Hereford Journal*, April 15, 1929].

Willie Joe Largent finally became an ardent lover, master breeder, and promoter of Polled Herefords. He judged many of the major Polled Hereford shows and garnered many champions on a national level.

An American Problem

The question "What to do with the polleds?" was an American problem for the first thirty-five years of the breed's existence, which had its birth and initial nurturing in America. The other Hereford societies around the world, being under a predominantly tradition-bound influence, tended to ignore Polled Herefords, obviously thinking that they would become extinct from lack of public interest.

The determination of the practical-minded proponents of polleds was underestimated by all. As they were tested, ridiculed, and belittled, the stronger the polleds became, until it was obvious that they would permeate every Hereford stronghold in the world.

The following article appeared in the *American Hereford Journal* in 1935. It gives an unusual insight to the attitudes of the breeders toward that segment. Polled Hereford breeders insisted upon being considered a segment and not a separate breed, but the traditionalists made it very difficult for many years to develop a spirit of unity within the breeders for a two-segment concept. Here is the article:

English Herd Book and Polled Herefords

The rapid growth in popularity of Polled Herefords in Canada, Australia, South Africa and New Zealand in recent years has raised the question of the attitude of the English Hereford Herd Book Society towards the muley Whitefaces. Up to the present, Polled Herefords have been denied admission to the English herd book, but as long as there were no Polled Herefords in the British Isles this made little difference.

With the establishment of herds of Polled Herefords in the British overseas dominions, and the use of American-bred Polled Hereford bulls on cows registered in the English Herd Book and the near approach of the time when, undoubtedly, Polled Herefords will be introduced into the British Isles themselves, the question assumes considerable importance. In the United States there has never been any question as to the eligibility of Polled Herefords to registration in the American Hereford herd book because of their lack of horns. The question of their eligibility to record rested entirely upon their ancestry. If their sires and dams were recorded, the animals were eligible to record regardless of whether they had horns or not.

The bull upon which the Double Standard Polled strain in the United States was founded, Giant 101740 (1), was a registered bull. When application for registry is made to the American Hereford Record it is accepted or rejected solely upon the ancestry of the animal, and not upon whether it is horned, scurred or polled.

On this subject, the noted English writer on agricultural and livestock, Harvester, in a recent issue of the *Hereford Times*, says: "For many years the Polled Hereford has been very popular in North America. This popularity of polled and dehorned Hereford cattle seems to be spreading to Canada, Australia, New Zealand and other parts of the world, and the question arises: How will it affect our own Hereford Herd Book Society, which only registers the horned type, so far as the export trade is concerned?

"The council of the Shorthorn Society of Great Britain and Ireland has considered a cable from the Shorthorn Society of Australia, asking what is the parent society's attitude towards the registration of Polled Shorthorns, and this society is reported to have decided not to recognize Polled Shorthorns for the purpose of registration. Is that to be the attitude of the Hereford breed society in the home of the foundation of the breed, or is the matter open for consideration?

"It requires thought. There are some people, even in England, Wales, Scotland and Ireland, who think that there is a future for the Polled Hereford on our own shores. In the meantime the inclination for the Polled Herefords overseas is extending. A breeder at Gisborne, New Zealand, writes: 'Polled Herefords are becoming increasingly popular in New Zealand, and now practically half of the stud Hereford breeders here are using polled bulls.' There have been importations in New Zealand and Australia from the U.S.A. With horned stud cows these Polled Hereford bulls are being used. Polled cattle of any breed are highly favored for yards and trucking, and they certainly look very attractive. It would be interesting to hear of the conclusions arrived at in Hereford" [*American Hereford Journal*, October 15, 1935].

6. Problems That Divided

The Double-Registration Dilemma

The Double-Registration Dilemma

The formative years of Polled Herefords were filled with all the problems typical of any new breed. There were a few problems, however, that were uncommon to the beginning of other breeds and were stumbling blocks for the first sixty years of the breed's history.

The most burdensome and tenacious problem blocking the progress of the breed and siphoning off the breed's resources that should have been used for promotion was "double recording." Many breeders felt that an endorsement by the American Hereford Association was necessary to validate Polled Herefords and make the records creditable, while the APHA must keep the records to maintain the record of polled ancestry.

For many years the AHA showed no desire to recognize the polled as a separate segment nor would it cooperate by providing polled identification on the certificates the association issued. Thus there was the need for a separate polled record maintained by those concerned with determining polled breeding. This need produced a constant and continuous debate and a long series of informal negotiations among breeders to eliminate the double-recording expense and burdensome paper work.

No doubt the expense and complexities, along with the added clerical work of double registration, discouraged the recording of some Polled Herefords for many years. Likewise, the revenue denied the association when breeders failed to register their cattle placed the APHA under severe financial stress and was a deterrent to promotion of the breed.

The letters that follow provide an insight into the controversy and problems of those early, formative years. They also present a portrayal of the character of the breeders of that day. One must appreciate their honest and straightforward attitudes as they expressed their opinions and wrestled with the problems that became almost unendurable at times.

The slow pace of progress must have been discouraging and no doubt many less determined breeders fell by the wayside and either quit the seed-stock business or moved to another cattle breed.

Since the birth of the breed, the problems involved in double registration had been a thorn in the side of Polled Hereford breeders. The duplicated paper work, along with the double registration costs, simmered as subjects of conversation during prosperous times but erupted into full-blown controversy during unprofitable times.

From 1920 to 1933, a period spanning thirteen years, the plight of the seed-stock producer had grown steadily worse. It was during the great depression of the mid-1930's that attention was focused on the inconsistency of keeping two sets of records. Numbers of calves registered dropped sharply and revenue to maintain the office staff and promote the breed was drastically reduced, forcing the curtailment of association activities. It was in the midst of times like these that supporting two associations to do the same work became a source of serious discussion.

Lewis Johnson, of Johnson Bros., Jacksboro, Texas, a pioneer breeder and one who became personally involved in advancing the cause of the breed, wrote on January 21, 1933, that finances were so low that no cash premiums could be given at the last show held the previous February and he raised a serious question about even attempting a show in 1933. He further emphasized the seriousness by pointing out that the treasury was behind $1200 on Gammon's salary and the net assets were only $450 above the current bills.

The cash position of the APHA, was the clear-cut result of breeders recording their calves in the AHA rather than the APHA. It was obvious that when breeders were in a financial bind they would prefer the AHA registration certificate and would forego the added expense of paying for the APHA certificate. (The APHA had required that all Polled Herefords be recorded in the AHA before accepting the applications for APHA entry.)

The untenable position of the APHA was disturbing to many Polled Hereford breeders and brought forth several suggestions for solutions. Special assessment of the breeders was suggested but not given much encouragement. Another suggestion was to completely withdraw from the AHA and require Polled Hereford breeders to have only the single APHA certificate.

A third alternative provided for the AHA to accept all applications and issue the single certificate. This scheme provided fifty cents from each registration to be given to the APHA for promotional purposes. Johnson summarized by stating that:

"More breeders have suggested that some sort of cooperation or consolidation be sought with the Kansas City Association [AHA], along the lines of the consolidation made some years back between the Shorthorn Association and the Polled Shorthorns. The amount of this consolidation to be determined and worked out later, if no more than to get the Kansas City Association to take over the record work, namely the recording and transferring. One feature of this plan has been suggested that the polled certificate be abandoned and only one certificate, namely the one now issued at Kansas City [by the AHA], be used but that the polled character of the animal be shown on said certificate."

There is no record in the annals of livestock history of an issue so fundamentally important to the advancement of a breed extending for so long a time. For sixty years the topic of conversation, wherever polled breeders met, dealt with the enigma of double registration and the financial burden imposed on the association and breeders alike. The primary concern of polled breeders was to maintain identity as a segment of the Hereford breed and have access to promotional funds to spread the advantages of the polled head.

Hugh White, of Keller, Texas, a prominent breeder, lawyer and, oil field operator, took a special interest in the breed's number one problem. In the following correspondence with Gammon he reveals a strong desire to maintain the polled identity but realizes the necessity for making some discretionary economic adjustments. It is interesting to observe that White's letter is dated May 20, 1935, and was received the following day, May 21, by Gammon and answered that same day. Today, postage cost has increased by 500 percent and delivery time by about the same proportion in spite of the advancement in speed and transportation.

Here is White's letter:

Mr. B. O. Gammon, Secretary   May 20, 1935
American Polled Hereford Association
Old Colony Bldg.
Des Moines, Iowa

My Dear Bert:
    What I am about to say will be a surprise to you, I know, but you are used to surprises, so this one will be just another incident.

John Lewis and Lewis Johnson have converted me to their way of thinking regarding Polled Hereford recording. I have been thinking about this matter ever since our discussion at Fort Worth in 1933, and the more I thought of the plan, the more I favored it. However, I held back on it until I was thoroughly convinced. The opposition to the so-called "Kinzer Plan" by such men as yourself, John Kelleher, Boyd Radford and Arch Dunbar, has, for some time, been my sole reason for opposing the plan. At first it seemed to me to be almost impossible to write a contract covering the matter, but after having made the attempt, it appears to me that a very simple contract will cover the matter. The vital clauses of the contract, according to my way of thinking, are:

1. That for a consideration the American Hereford Association will record Polled Herefords on a certificate bearing a picture of a polled head, that the letter 'P' will be either suffixed or affixed to the official number and made a part thereof, so that such number will designate polled animals in a pedigree as does the letter "X" in the number of a Polled Shorthorn.

2. That the present records of the Polled Hereford shall forever remain the property of the Polled Hereford Association, but that same shall be loaned to the American Hereford Association for the duration of this contract and all continuations thereof.

3. That either association can withdraw from said contract upon proper notice, said notice to be given in writing, and two days in advance of anniversary of contract, otherwise contract to remain in full force and effect.

4. American Hereford Association to make the following charge for such certificate of registration; the same fee as charged for recording horned animals, this amount to be the property of the American Hereford Association, and in addition thereto the American Hereford Association will charge an amount designated by the Directors of the Polled Hereford Association, the latter amount to be paid into the treasury of the American Polled

Hereford Breeders' Association, and to be spent according to the direction of the Board of Directors of the American Polled Hereford Association, said amount to be paid monthly.

5. The American Hereford Association agrees to furnish to the Secretary of the American Polled Hereford Association, free office space in its building at 300 W. 11th Street, Kansas City, Mo., but it shall not be obligatory for the Polled Hereford Association to avail itself of this office.

The contract, of course, would be written more carefully and could embrace other clauses that I may have overlooked, but it seems to me that the contract could be simple and would not burn the bridges behind us, as we could reenter the recording field at any anniversary date.

You, of all people, are entitled to know what my position is and I feel that in telling you what that position is, is telling all concerned since the secretary is the fountain head of information. This is my purpose in writing this letter. If I am making a mistake, it is one of the head and not of the heart.

In the meantime, I will be quite as vigilant as in the past in securing new members and in securing cattle for record, because we will need new members as badly after the new plan has gone into effect as now and polled identity for polled cattle is as important as ever.

The *Journal* came today and I have read the report of the Association meeting and of the sale with a great deal of interest. I note that the National Polled Hereford Show will be held in Des Moines in 1936. Is it contemplated that any cash premiums will be provided for?

Hoping to hear from you at your earliest convenience, I am

> Sincerely yours,
> Hugh H. White

The following letter is from B. O. Gammon, Executive Secretary of the APHA, in reply to Hugh H. White, regarding registration of Polled Herefords by the AHA.

Mr. Hugh White   May 21, 1935
Keller, Texas

Friend Hugh:

Your letter of the 20th was, of course, as you surmised it would be, a distinct surprise to me, and since it has been in my possession only a few minutes, I do not feel yet quite ready to reply further than with this acknowledgment. One provision of the contract which does not appear in your suggestions and which I deem to be of vital importance is a provision which occurs in the requirements of registration of Polled Shorthorns about which I visited with Mr. Harding on last Saturday. That provision is a requirement that no polled animal will be recorded except as a polled animal. This you will note would effectively prevent breeders of Polled Herefords from recording them only in the American Hereford Record and avoiding thereby the payment of the extra fee which is to go to support promotion work for polled cattle.

This provision it seems to me is absolutely vital to the future of Polled Herefords because without it a large proportion, perhaps all, our breeders of Polled Herefords would say that they are not interested in spending any additional money to support promotion work and since they do not have to spend it they will not do so, thus leaving us in much worse condition than we have ever been in the past by reason of refusal of breeders of Polled Herefords to patronize this Record.

Mr. Harding tells me that so far as they know they are not recording any polled cattle simply as Shorthorns. They use a different color and form of application blank for polled cattle and require that all polled animals shall be applied for on this particular form and color of blank.

Another provision, perhaps meant to be covered in your paragraph number 2, would be that the Polled Hereford Association should have an equal and independent ownership of all entries of polled cattle made in the AHR after the herd books are combined so that should we desire to withdraw we would have ownership and possession of all entries added to the Polled Hereford Record during the period of the combined herd books. I think perhaps this was in your mind in writing of paragraph number 2.

I certainly appreciate the frank and "above board" attitude you take in this matter in publicly stating your views to me with the understanding that they may be passed on to other interested parties. In this way no one can take exception to your expressing your opinions nor can they charge you with an attempt to secretly undermine the association and while many of them may differ with you in their judgment as to the wisdom of the move, they certainly cannot have anything but the highest respect for the way in which you bring your views before the fraternity.

With personal regards, I am

Yours truly,
B. O. Gammon

This represents only a part of the correspondence covering the first serious attempt by Polled Hereford breeders to eliminate double registration.

The plan proposed in 1935 and rejected by the AHA was essentially the same as the one adopted in 1962 which was known as the single joint certificate contract. The points that prevented complete agreement were the same that occurred fifteen years later during the second attempt at unifying the breed—a formula for collecting funds and an agreement that Polled Hereford records belonged to Polled Hereford breeders. In essence, polled breeders were apprehensive about a complete take-over by a horned board and loss of identity and the right to promote their own interests.

The following memorandum and resolution tell the story quite clearly of what happened fifteen years later when a refined version was finally discussed by the boards representing the two associations.

Resolution

That the Resolution of the Board of Directors of the American Polled Hereford Association of date November 1, 1948 unanimously passed reading:

"That we are not in favor of turning the American Polled Hereford Records over to the American Hereford Association in order for the American Hereford Association to issue one certificate; and suggest to the membership that we either continue to issue certificates as we are doing, for the time being, or if the membership wants one certificate, that it be issued by the American Polled Hereford Association."

Be and is reaffirmed, and the Committee appointed to confer with the American Hereford Association on the question of issuance of a single joint certificate be instructed and directed to advise the American Hereford Association that the American Polled Hereford Association is in favor of retaining the American Polled Hereford Association registration number, and that the American Polled Hereford Association issue the joint certificate, and that the Committee be instructed to ascertain whether or not the American Hereford Association will cooperate in the issuance of a single joint certificate on that basis.

Passed by Board of Directors of American Polled Hereford Association unanimously on November 6, 1949.

Confirmed and approved by motion adopted by Membership at Memphis Meeting on November 8, 1949.

ATTEST: D. W. Chittenden, Secretary, American Polled Hereford Association.

Memorandum

From: Officers & Directors of the American Polled Hereford Assoc.

To: Members of the American Polled Hereford Association

Subject: Single Certificate

With all Officers and Directors assembled at the Hotel Muehlebach, Kansas City, Missouri, 9 A.M., August 7, 1950.

Concerning the report of the Committee (Adna R. Johnson, Chairman, and M. P. Moore), upon the question of single registration.

Recognition of American Polled Hereford Certificate

The first effort on the part of this Committee was to determine whether or not the American Hereford Association would recognize the certificate of the American Polled Hereford Association.

Upon motion duly made, seconded and unanimously carried, the Secretary of A.H.A. was instructed to apply the rules of the American Hereford Association for both horned and polled cattle. The answer, therefore, was in the negative.

Joint Certificate

Our next effort was to secure the issuance of a Joint Certificate by the American Polled Hereford Association and the American Hereford Association. To this end several meetings were held with representatives of the American Hereford Association, and a number of letters exchanged over a period of the last year and a half.

We regret to report that the suggestions of the American Hereford Association for a joint certificate for polled breeders as set forth in the above correspondence prohibits acceptance by your Board of Directors.

The question next discussed by your Board of Directors; whether or not Polled Hereford Association members would desire to consider the question of eliminating the present requirement of an American Hereford Association certificate before registration in their own Association. This would be a matter for the determination of each breeder and would not in any manner interfere with, or affect, Polled Hereford breeders who desire to register their cattle in the American Hereford Association. The American Polled Hereford Association has neither the authority or desire to recommend to individual breeders their action. Further consideration of the Board upon this question was deferred to the Director's meeting of August 7, 1950, when it was decided that the factual information be presented to the members of the American Polled Hereford Association for their consideration (please see attached correspondence).

We believe that any further procedure upon this question will depend upon the wishes of our membership. At the present time, there is some difference of opinion between our members, and in a matter of this importance, your Directors recommend that no further action be taken in this matter at this time. We are of the opinion that time and future development of our cattle will clarify this situation, and we suggest that you give appropriate thought and consideration to this important question. To this end, the directors have ordered and authorized the appointment of a continuing committee whose task will be to keep fully acquainted with the situation as it develops and keep the membership advised.

John M. Lewis, President
D. W. Chittenden, Exec. Secy.

7. Trying Times

From the Pinnacle to the Pit
The Dwarfism Problem

From the Pinnacle to the Pit

The year 1953 was a banner year in the history of Herefords. Over 500,000 registrations for both horned and polled segments were issued. In October, 1953, the President of the United States, Dwight D. Eisenhower, dedicated the American Hereford Association's building.

In 1953 the greatest array of cattle ever assembled in a show ring appeared in the Hereford show at the International in Chicago. Herefords made up 68 per cent of the annual total registrations of the three major beef breeds. Truly the breed had reached the pinnacle.

In 1965, twelve years later, the AHA recorded 302,000 annual registrations, and by 1972 registrations totaled only 236,000. For those with more than a passing interest in Hereford heritage this continued downturn in numbers is enough to shock one into reality. A twenty-year trend of such consequence must be interpreted as significant and of more than just seasonal or climatic cause. It is reminiscent of the ten-year span from 1920 to 1930, when the once dominant Shorthorn breed dropped in numbers from 416,980 to 260,312 and finally to 35,000 annual recordings in 1973.

The following chart, appearing in a 1930 edition of the *American Hereford Journal*, indicates the number of registered beef cattle, by states, of the three major breeds in 1920 and 1930.

REGISTERED BEEF CATTLE BY STATES
Census of 1920 and 1930.
(Polled Herefords included with Herefords, and Polled Shorthorns with Shorthorns.)

	Herefords.		Shorthorns.		Angus.	
	1930	1920	1930	1920	1930	1920
Alabama	1,161	1,524	182	1,508	514	883
Arizona	8,069	2,023	98	224		
Arkansas	1,297	1,726	431	2,128	137	893
California	9,890	5,238	5,198	3,045	1,406	1,182
Colorado	23,451	17,270	6,974	6,989	974	615
Connecticut	130	138	251	207	26	71
Delaware	182		3			
Florida	80	687	8	96	165	577
Georgia	1,088	1,799	72	1,293	277	497
Idaho	5,559	5,464	4,110	6,298	84	295
Illinois	9,279	16,370	16,407	39,093	3,862	10,106
Indiana	2,873	6,615	6,893	16,147	1,654	4,807
Iowa	29,555	40,894	38,230	69,560	13,364	27,457
Kansas	43,813	38,695	24,846	27,404	2,673	4,700
Kentucky	4,405	4,375	2,395	3,536	1,908	1,706
Louisiana	524	1,340	191	840	120	313
Maryland	148	106	392	414	167	135
Massachusetts	282	344	409	717	61	112
Michigan	1,957	1,825	5,783	11,712	696	1,519
Maine	1,338	1,530	1,167	442	585	63
Minnesota	6,082	10,787	23,592	32,419	2,072	5,398
Mississippi	1,023	2,640	100	1,580	199	2,072
Missouri	30,087	32,609	16,073	30,517	8,397	12,916
Montana	18,811	10,699	8,194	5,621	1,155	927
Nebraska	43,350	27,418	19,280	32,777	2,692	4,640
Nevada	2,399	2,422	978	617	5	15
New Hampshire	728	888	650	676	51	
New Jersey	6	2	14	71	28	12
New Mexico	18,932	14,563	140	434	72	111
New York	460	60	1,451	1,118	1,459	248
North Carolina	820	933	389	732	387	786
North Dakota	8,159	7,089	15,851	14,723	2,159	3,124
Ohio	2,901	3,229	7,200	17,324	1,160	2,642
Oklahoma	14,015	12,133	9,367	22,019	1,104	1,876
Oregon	5,689	4,182	4,227	3,758	639	776
Pennsylvania	1,044	686	2,303	3,676	298	428
Rhode Island	7	64	26	4	9	
South Carolina	236	862	3	103	84	328
South Dakota	17,583	21,663	11,803	23,293	1,822	4,788
Tennessee	3,166	4,084	1,611	3,286	1,571	3,799
Texas	94,951	70,021	4,441	4,371	948	2,605
Utah	5,118	5,978	1,772	2,007	58	62
Vermont	206	371	403	491	112	28
Virginia	2,365	2,135	2,492	4,289	855	1,121
Washington	1,189	935	3,941	3,281	641	265
West Virginia	5,550	4,118	1,583	1,709	1,278	1,962
Wisconsin	943	2,203	7,295	13,125	617	1,539
Wyoming	17,772	11,845	1,293	1,305	175	115
Total	448,767	405,582	260,312	416,980	58,715	108,512

There developed over a period from about 1950 to 1960 a public resistance to Herefords. Some felt that the reason might be public condemnation of horns. Others thought that it was a lack of promotional outlay or that there were problems unique to Herefords. However, while Herefords were taking the steady downward trend in numbers, other breeds subjected to the same problems, the same climatic conditions, and the same market trends continued to progress in numbers and public acceptance. From 1950 to 1970, Angus climbed from 22 to 47 per cent of the total annual registrations of the three major breeds. Polled Herefords progressed from 12 to 20 per cent. During the same period horned Herefords dropped from 50 to 33 per cent of the total.

The Dwarfism Problem

The first century of beef production in the United States presented many problems and challenges to the seed-stock producer. Drought, pestilence, and economic disaster were traumas he learned to take in stride. The less hardy, the inner-and-outer, would drop by the wayside, but the grass-roots cattleman, the professional, would endure these tragedies and survive through sheer grit and determination.

There was one malady, however, that brought chaos and panic. It came in the form of cute little bug-eyed calves that fascinated the uninitiated, some of whom even thought that they had been blessed with a genetic bonanza. The term "dwarfism" was later to bring fear and panic to the ranchers, who had learned to handle most any other kind of problem.

Although these odd little calves had been known to occur in almost all breeds in generations past, their frequency was low enough not to create great concern or economic impact.

The incidence of dwarfism did not occur at an alarming rate until the middle-to-late 1940's. A few occurred in some herds, mainly in the Middle West and Rocky Mountain areas. Several theories were advanced at the outset by nutritionists and geneticists. Cattlemen were at a loss when it came to hazarding a guess about the cause of the phenomenon. Mineral supplements were rushed to the market with claims to be the antidotes. Special fertilizers were recommended during the frustrating period of research.

A tribute must be given to members of the American Hereford Association and their leaders for the candid and objective way they approached the problem of diagnosis. While other breed associations had the same problem, their methods of researching the cause were much more subtle and secretive. There was a great reluctance to admit that the problem existed within their breeds, an attitude that contributed very little to identifying the cause of dwarfism.

The incidence was finally traced to specific family lines and ultimately to key bulls and cows, many of which were long since dead. Through a survey of the breeders a fairly reliable bank of information was collected, which was used to develop the celebrated and widely publicized pedigree-check program. Its basic design was to be used as a research tool for owners to check the genetic soundness of the animals in their ownership to determine ancestral freedom from dwarfism.

It is not my purpose here to deal with the genetics of dwarfism but simply to record the historical aspect. The pedigree-check program was basically a sound idea, but in order to serve the industry in a constructive way, it should have been a companion program with the progeny test. Unfortunately, the progeny test, although used by several breeders, never gained the favor and support of the breed associations. Consequently, breeders who were fortunate to have so-called clean lines of pedigrees discounted the progeny test and continued to refer to "dirty ancestral lines." The pedigree check deteriorated into a marketing tool for those with clean lines, or, rather, lines of pedigrees that had never been reported as having had dwarfism problems. A breeder that had a "clean pedigree report" from the American Hereford

Association had an automatic license to bill his sale as a "clean sale." An example is a man in the Hereford business who is enjoying a profitable venture. He has a good productive herd and is selling range bulls for enough to make a decent living. Unknown to him, another breeder had several generations earlier developed some of the same family lines, in which a bull with abnormal genes was used. This breeder, now having dispersed, decides to report the fact that the bull he used had sired a dwarf or that a descendant of the bull was an offender. Now the current breeder who is enjoying a profitable business receives the report or "whisper." He will react one of two ways. If he is financially able to take the purest attitude and establish an open reputation of being totally dedicated, he may consign all his cattle to a commercial herd and report his sanctimonious action to the public. On the other hand, if his life savings, his standard of living and the education of his children are at stake, he may choose the more practical approach. He will then set about to organize a dispersal sale, scatter his cattle, most of which and perhaps all of which are "clean" but dirty only by association, and eventually confess his problem, contributing another blow toward wrecking the breed.

Why has a major competing breed continued its upward growth in numbers while being subject to the same problems? It is very simple. Breeders let the offenders, "bulls that dwarf out," fall through the screen, eliminate them, and keep the good individuals for breed progress, while the Hereford people jerk the trap door from under an entire bloodline or family, and in effect an entire herd goes down the drain. Not many breeders have the resources to absorb such shock, and time has proved its effect upon an entire breed. Valid questions arise: Why condemn a family because there is one offending relative? Why convict so many innocent in order to cull only one offender?

At a time when beef-cattle numbers were greater than at any

time in history and the need for even greater production was badly needed, why the great exodus from the Hereford business? The answer should not surprise us. Herefords promoted their liabilities, while the competition promoted their assets. The entire breed, both horned and polled, were victims of "whisper whipping," which is a by-product of an antiquated program designed to be effective as a research tool but is misused as a marketing tool.

Geneticists identifying the presence of at least forty-eight known abnormal traits admit that fewer undesirables have been identified in Polled Herefords than in any of the other major breeds. That Polled Herefords have less lethal genetic trash in their makeup and more of the basic money-making traits is an asset that should be jealously protected and promoted with zeal.

Every effort should be exerted to prevent the introduction of germ plasm into Polled Herefords, which may result in scattering genetic trash. Continuing through progeny test to identify and eliminate the genetic trash that may exist while at the same time locating the superior-performing individuals is a must for breed progress. The breed's Superior Sire Program referred to later is the key to breed progress and is the best insurance against a recurrence of this traumatic era.

Following a period of intensive research and study of the dwarfism problem, the thinking of the breed's leaders began to crystallize, and a definite program was developed. First, it was obvious and agreed to by geneticists and cattlemen alike that no breed would ever be completely free from some genetic abnormalities. Second, there was a virtual "halo" above the head of a bull or cow that had an endorsement by a "clean-pedigree check," when in fact this was no guarantee of genetic freedom from dwarfism but only that none had been reported. Third, if dwarfism was to be reported and controlled, the halo must be removed from above the pedigree-check and dwarf-report program and more emphasis placed on the progeny test. Fourth, the pedigree check should not be used as a merchandising tool but only in the actual process of tracing the source from which problems occur.

As stated in the Recommended Terms & Guarantees of the APHA:

"Polled Herefords have been most fortunate in being relatively free of inherited defects which would impair the breed's ability to perform and reproduce. Some recognize the useful genetic make-up of Polled Herefords as having come from the parent stock of original Polled Herefords. These cattle had a generous measure of the blood of the Hereford breed's foundation animals that were recognized as free from inherited defects and with a preponderance of growth potential. Evidence of the genetic soundness of Polled Herefords may be recognized in that there has not been a Polled Hereford reported as having produced a dwarf that traces in all lines to the original Polled Hereford foundation stock.

"In 1968 the directors of APHA completed a two-year study to determine the frequency of genetic problems for inherited defects. It has been determined that dwarfism, as such, is of such low frequency that it no longer exists as an economic problem in Polled Herefords. However, it is believed that an intelligent industry will be constantly on the alert and take such precautions as are necessary to protect the breed from unforeseen genetic problems. It is necessary, also, at the same time, to build and maintain public confidence in the breed and avoid any practices that may create fear on the part of new breeders and discourage them from entering the Polled Hereford industry.

"The American Polled Hereford Association has the capabilities of pedigree analysis on dwarf producers. However, the Advisory Committee on Education and Research has recommended that the use of this information be limited to research. It further recommended that pedigree analysis not be used as a merchandising tool. It is recognized further that emphasis must be placed on all traits of economic importance and dwarfism should be classified along with all other forms of genetic trash and similar means of detection and control be used. After thorough study and research, the Board of Directors of the American Polled Hereford Association has made the following recommendations.

"That the American Polled Hereford Association:

1. Expand the current Association recommended terms of sale to cover specific inherited defects.

2. A comprehensive educational program to be implemented to encourage breeders to promote confidence in the genetic soundness and fertility of their cattle by the use of the recommended guarantees in making sales.

3. Encourage emphasis on breeding programs to prove superior sires in performance, reproductive efficiency and freedom from genetic trash.

4. Maintain the current research file on all Polled Herefords which have produced identifiable genetic abnormalities.

"That Polled Hereford breeders:

1. Continue to build confidence in the individual breeder by his assuming the responsibility for the genetic soundness of the cattle he sells as he does now regarding breeding soundness and freedom from other defects.

2. Discontinue the practice of using the pedigree check as a merchandising tool.

3. Continue to inform the American Polled Hereford Association Education & Research Department concerning abnormal calves, i.e. dwarfs, etc. This information to be used for research purposes only and to help in keeping the Association's files current concerning dwarfism and other abnormalities that may arise.

4. Utilize the recommended terms and guarantees by the use of the following statement in catalogs of private and consignment

sales: The cattle offered for sale are sold under the terms and guarantees recommended by the American Polled Hereford Association with reference to health, freedom from defects, reproductive unsoundness, and freedom from specific genetic unsoundness. The recommended refunds or adjustments will be made in case animals fail to live up to seller's guarantee."

8. Family Feuds

Domestic Unrest
The Merger for Five Years
The Split

Family Feuds

Domestic Unrest

A restlessness was growing throughout the Polled Hereford segment. It was brought on in part by the independent attitude of the American Hereford Association and its paternalism over not only the Hereford scene but the entire beef cattle industry. For thirty years or more the AHA had been unchallenged from the standpoint of industry supremacy and leadership. Angus cattle were growing rapidly in numbers, however, and an aggressive promotional program carried on by the Angus Association was challenging the numbers and the leadership of the AHA.

The rift between the two segments of the Hereford breed was becoming more obvious and was beginning to be a thorn in the side of the AHA as it began to compare its total numbers with the growing number of Angus. After about 1958, when Polled Herefords recorded more than 100,000 head, there grew a tendency on the part of aggressive Polled Hereford breeders to challenge the leadership and the dominance of the AHA. In 1957 a group of Polled Hereford breeders met for the express purpose of encouraging the entire Polled Hereford segment to single-record in the APHA. As this movement began to grow, the AHA threw its superior field force and promotional machinery into the battle and succeeded in dividing Polled Hereford breeders down the middle. It now became obvious that, if the withdrawal attempt on the part of Polled Hereford breeders was successful, it would only succeed in carrying with it less than half of the Polled Hereford breeders. That would leave at least one-half and probably more to remain with the AHA.

The AHA had influenced many in its planned attempt to convince Hereford breeders throughout the land that the APHA records and registration certificates were invalid insofar as acceptance by members of the World Hereford Council was concerned. Because of the AHA's desire to maintain the one-breed image, a series of meetings was initiated to develop a single joint certificate that would be acceptable to both segments of the Hereford breed.

In 1947, B. O. "Bert" Gammon, who until that time was the only full-time secretary the APHA had ever had, was seventy years old. It was decided by the Board of Directors of the APHA to employ a field secretary who could do most of the traveling and relieve Gammon of many of these burdensome responsibilities. D. W. "Don" Chittenden was employed in this capacity, and within a short time Gammon retired, to become secretary emeritus, and Chittenden was elevated to executive secretary.

Chittenden, as a field man for other publications, and especially for the *American Hereford Journal*, was conditioned to have strong personal feelings about the segments. He approached his job with enthusiasm and aggressiveness. His strong will, determination, and disregard for the art of compromise created a virtual whirlwind of turmoil wherever he went. His outspoken methods further tended to draw the lines of differences among breeders within the Polled Hereford segment and further alienated breeders of other breeds. Needless to say, a man of lesser talent and aggressiveness might have been completely defeated by the formidable competitor, the AHA and the horned breeders.

By 1960 the course had been established and the battle lines drawn, and Chittenden became more and more closely identified and allied with those who were independent-minded and were ready and willing to withdraw from the Hereford breed regardless of the cost. As these differences began to affect the attitudes of the directors of the APHA, the members of the board began to align themselves on one or the other side of the battle lines. A majority of the board supported the proposal for a single joint certificate and an agreement with the AHA. A minority of the group, along with the executive secretary and the president, set themselves solidly against any agreement, even if it meant splitting the Polled Hereford segment down the middle.

In the summer of 1962 an agreement was reached between the majority of the directors of the APHA and the directors of the AHA for entering the single joint certificate. The agreement was to be presented to the memberships of both associations in the fall of 1962. The AHA, at its annual membership meeting in October, ratified and accepted the agreement between the two associations.

However, at about the same time, at an APHA board meeting, an injunction was filed by the minority group on the board that prevented the APHA from entering into agreement with the AHA. It was during this meeting that the APHA board asked for the resignation of Chittenden, and he was terminated October 16, 1962. L. J. "Jim" Harris was appointed interim secretary.

It was a tense period when the rumor mills were working overtime and the atmosphere was charged with emotion and reaction. For the next few months the conversation around cattle circles was dominated by accusations, rumors, and assumptions. There was a buildup of emotion and tensions that

D. W. "Don" Chittenden, APHA Executive Secretary, 1947-62.

resulted in the most memorable meeting in the breed's history. Breeders came from far and wide to Jackson, Mississippi, to attend the National Show and Annual Meeting, December 6, 1962. They aligned themselves on one side or the other of this controversial issue, which was to set the course of history for the breed.

One group said that the agreement was a sellout to the AHA, a giveaway of the birthright of Polled Herefords, that the APHA would cease to exist, and that Polled Herefords, even as a segment, would become extinct. The other group said that it was simply a plan to relieve breeders of Polled Herefords of the burden of double recording and maintaining two registration certificates.

History has proved since then that both positions were right to a degree. In retrospect it can be said that the Polled Hereford industry was gambling everything that it had and placing itself in an untenable position in order to maintain unity among its breeders. It was virtually going into captivity as a unit, hoping that sometime in the future it would emerge as a strong, united, and free organization. Debate was hot and uninhibited, and when the smoke cleared away, it was obvious that at least three-fourths of the breeders were for signing the contract and entering into a program under which they could eliminate double recording forever.

The Merger for Five Years

It was not until some time later that the real impact of the provisions of the contract began to be felt by the APHA. In essence, the association had traded more than half of its gross income and taken on the added burden of providing all of the promotion for Polled Herefords throughout the land. In the past the AHA had been providing some Polled Hereford promotion. The APHA found itself in the position of having to provide more service and promotion for its breeders and less funds to do it with. It became apparent that the powerful AHA could sweep the entire Polled Hereford breedership into its pocket at the end of the five-year contract. The APHA could only look forward to a period of time when greater demands would be placed upon it to provide breeders of Polled Herefords with the same level of service that the AHA was financially able to give its horned breeders. The great disadvantage was obvious. Through the years of double recording the AHA had been able to build great reserves and would continue to be subsidized by the heavy tax paid by Polled Hereford breeders in order to maintain unity within their ranks. There were serious prospects that all polled breeders would become disenchanted because of the slender budget remaining to promote Polled Herefords and want to cancel the entire agreement and abdicate to the horned segment.

Not to be overlooked is the fact that, after the five-year period during which the AHA kept all the records for Polled Herefords, the APHA would find it most difficult to reconstruct its records for re-establishing a record-keeping system. Obviously the American Polled Hereford Association would find itself completely at the mercy of the American Hereford Association in renegotiating its contract at the end of five years.

A complete assessment was made of the APHA's position after three months of the contract period. It was obvious that more funds would have to be made available to build a field staff, develop a promotional program, pay premiums at fairs and shows and, in general, serve the breeders of Polled Herefords.

In April, 1963, in their regular spring board meeting, the directors voted to raise the registration fees from $2.50 to $3.50. This provided the needed revenue to sustain the APHA and allow it to maintain the records which were vital for the eventual reconstruction and resumption of recording service five years later.

The field staff provided an important means of communicating with the breeders about the progress of the contract arrangements, in addition to their routine field service.

The Split

As the contract renewal date of 1967 approached, the tempo of competitive activities picked up, and promotional conflicts surfaced more frequently. A typical example was the booklet, *Answers to 29 Basic Questions About Raising Beef Cattle*, published by the AHA. It contained derogatory statements, implications, and innuendoes discrediting Polled Herefords. It cast serious doubts on their genetic origin. It was plain now that Madison Avenue type advertising had entered the pedigreed-cattle industry.

Below are excerpts from the booklet, published in 1967:

"Why Horned Hereford bulls?

"Again, nothing succeeds like success and there is more success, more millions of research and development, more up-breeding to the modern, efficient beef conformation type and more people think so—than any other beef breed in all the world.

"The horned Hereford bull, because of his innate masculinity, will pick out his harem, even on the range, and stay with them. Did you ever see horned Hereford bulls bunched up around a water hole or several of them following one cow? The horned Hereford knows his job—no strange ideas here. And no problems from outcross mutations.

"It is completely true that the best Hereford bull you can afford doesn't really cost. He pays and pays—is capable of reproducing himself in kind 300, 400, even 500 times.

"Do you know any investment that will beat that for capital gain?

"Why do the true Herefords have horns?

"That's like asking, why does a real 'he-man' have whiskers. The horns on gentle Herefords are nature's trade-marks of masculine vigor, strength and self reliant capability in taking care of themselves and protecting the herd under all conditions. The horns also establish the status of the bull in the herd. No mistake about it—he is boss of his harem against all comers.

"While the mysteries of genetics are not yet fully understood, it is a truism of breeding hybrids and mutations, that almost invariably, when you breed for one trait you lose other profitable characteristics. That is why the horned Hereford does not suffer from genetic weaknesses common to many outcross mutations. He is pure bred. Certainly he is not a hybrid and certainly he is not 'mixed up' with uncertain future genetic tendencies.

"The true horned Hereford bloodlines, constantly improved with blood of the best individuals of the same breed for scores of generations, have that much more experience and success behind them than later mutations. Truly, this is the great 'middle road' that leads to one goal—the production of more pounds of premium beef in the right places, faster, more efficiently and with more profit . . . and under all kinds of conditions."

This publication, implying that the polled breed was not "pure," so inflamed the tempers of Polled Hereford breeders and aroused their emotions that many were reacting and demanding that the APHA relationship with the AHA be reviewed. Many felt that the AHA was profiting on registration fees from Polled Hereford breeders and using the proceeds to finance promotional programs to malign the cattle that made the funds available.

Late in 1966 a series of meetings was held between the liaison committees of the two associations in an attempt to negotiate a workable new contract and continue the joint relationship. The APHA was paying the AHA about $30,000 each month for issuing the certificates and maintaining the ancestry files on the computer. It was necessary, however, for the APHA to maintain a duplicate record in case it should become necessary to reinstate its registry service to Polled Hereford breeders. All efforts to arrive at a compromise on fees and service charges failed. It was ironic when a staff member of the AHA, in reviewing a proposal that a data-service firm had made to do the APHA's work at about one-half the costs, said, "They will make a healthy profit off of you at that."

A final meeting between the two liaison committees (consisting of R. E. "Pat" Connolly, Bruce Purdy, Lynn Storm, Walter Lewis and myself, representing the APHA; and Henry Matthiessen, Jr., Marshall Sellman, Paul Swaffar, and Lorin Duemeland, representing the AHA) was called by the APHA board in May, 1967, in a last attempt to reconcile differences and arrive at a workable agreement. The AHA refused to compromise and, in fact, even requested an increase in the fees to be paid them.

The final straw, however, was the attitude of the AHA committee members toward the future use of the *29 Questions* booklet. When asked the AHA's plan for its future use, the answer was, "If there is a contract between the two associations, the booklet would remain in circulation but in the absence of a contract the booklet would be withdrawn."

An interesting observation was made by Lynn Storm, Dripping Springs, Texas, a polled representative on the committee, when he asked Henry Matthiessen, President of the AHA, "It would appear that the contract creates an atmosphere which encourages the use of derogatory promotional statements. Would you agree?" Matthiessen answered, "Mr. Storm, in the light in which you present it,

I would have to agree." On this note the APHA decided that any further efforts were futile.

There was an air of confident optimism on the part of the horned representatives, however, as they left the conference. They gave the distinct impression that they still felt the APHA would not terminate but would continue the contract. It was not until months later, at the termination of the joint processing, when the APHA was in the process of reconstructing the records the AHA had provided that the truth came out. The AHA felt secure in the belief that the APHA could never reconstruct the records into a workable system with the copies of the five years of pedigree recordings the AHA supplied. The gaps in the ownership records provided by the computer tapes were such that the association had to refer to file copies and rely on breeder records for an indefinite period.

If there was ever a time in the annals of a livestock-registry association when Winston Churchill's wartime statement about the Royal Air Force would apply, it certainly did at this time to the staff of the APHA. Churchill said, "Never before have so many owed so much to so few." It was only through self-sacrifice, dedication and determination on the part of the APHA staff that records were reconstructed and the value of Polled Herefords was maintained.

Below is the letter terminating the contract between the APHA and the AHA:

Board of Directors   April 19, 1967
American Hereford Association
Hereford Drive
Kansas City, Mo. 64105
Gentlemen:

On January 1, 1963, the American Polled Hereford Association, in good faith, entered into a contract with the American Hereford Association which provided: (1) that both segments recognize

ethics in promotion and condemned advertising which would
serve to align one segment, horned or polled, against the other; (2)
for a single joint certificate for Polled Herefords, and (3) a fee for
processing Polled Hereford work based on the expected volume at
that time.

Now we have progressed for 4½ years and the contract provided
that at the end of this period it could be reviewed, amended and
continued or terminated. In May of 1966 the American Polled
Hereford Association Board of Directors initiated a series of
meetings with the American Hereford Association in hopes that
the present contract could be amended to correct certain conditions which were creating a hardship on the Polled Hereford
segment of the industry.

The recent distribution of literature which is derogatory
toward the Polled segment, such as your recent pamphlet entitled,
"Answers to 29 Questions," (question 22) which states "Why Do
True Herefords Have Horns?" and implying that Polled Herefords
are not true Herefords, is considered a violation of Article X of the
joint certificate agreement. It is also considered a violation of Item
5 in the World Hereford Conference Charter which was adopted
in the 1964 World Hereford Conference which states: "To
maintain a high standard of ethics in all promotional activities,
both at individual and association levels." Your stated intention
to continue this type of advertising if the contract is renewed is
extremely detrimental to the Polled Hereford segment and
precludes a harmonious relationship as is the intent of the
contract.

The excessive fees charged Polled Hereford breeders for processing their work and the profits therefrom being used in the
promotion of horned Herefords is extremely unfair. A study has
revealed a savings to Polled Hereford breeders of more than
$150,000.00 a year by installing our own system which will result
in more adequate service to Polled Hereford breeders.

Polled Hereford breeders have felt there was a lack of Polled Hereford sponsorship in open shows in spite of their contributing more than $129,000.00 last year to be paid as premiums in these shows and other activities by virtue of penalty fees paid by Polled Hereford breeders to AHA. Polled Hereford breeders received less than $32,000.00 of these premiums in return in open shows last year.

The lack of a Polled Hereford voice in the sponsorship of open shows has created a desire on the part of Polled Hereford breeders to have more separate shows which would require financing by the APHA. Polled Hereford breeders feel that it is imperative that money collected from them should be used to continue the promotion and advancement of Polled Herefords undertaken by the APHA during the last few years.

We regret that after a series of joint meetings between the two associations and our repeated attempts to convey to you the unfairness to Polled Hereford breeders in continuing to subsidize the promotion of horned Herefords that you remain unyielding and uncompromising on the points at issue. More especially is this inconsistency pointed out when funds from Polled Hereford breeders are used in horned Hereford promotion which contains untrue statements by implication about Polled Herefords.

Realizing our strong responsibility to Polled Hereford breeders in protecting their interest, and recognizing it is no longer necessary for Polled Hereford breeders to double record because of the worldwide acceptance of APHA certificates, and after a long, careful study and with great reluctance upon our part, it is the unanimous consent of the Board of Directors with the full support of the members at their annual membership meeting, that we hereby notify you that effective January 1, 1968, the above described contract with its specific provision of joint processing of Polled Hereford work will be terminated pursuant to Article VII.

We appreciate the spirit in which you entered into the series of meetings which resulted in discussion on a high plane and a friendly and congenial atmosphere. It is our hope and desire that we can continue to work together in other areas in the future as in the past in solving problems and sponsoring events of mutual interest and benefit.

Sincerely yours,
R. E. Connolly
President

Matthew Heartney, Jr.   *Jim Harris*   *Temple Wells*

Finis McFarland   *Marcine Reagan*   *Ken Harwell*

Special tribute is paid to these tireless APHA associates who are representative of the many others whose dedication and efforts safely guided the Association through uncertain times.

9. Struggle for Identity

Origin of World Hereford Council

The first gathering of Hereford breeders that resulted in what is now known as the World Hereford Council took place from July 10 to 13, 1951. It was a forum designed primarily to discuss problems within the Hereford breed. The main topics of discussion were the importance of type, animal-disease control, and, probably the most significant and controversial of all the topics, *how to resolve the registry question and reciprocal exchange of records.* One of the unscheduled but most pertinent questions was "What to do with the 'polled problem?'"

The following account of this meeting and the quotations by certain representatives should prove interesting to beef producers everywhere and particularly to Polled Hereford breeders around the world.

Account of 1951 Conference

Mr. Schofield (Argentina) defined the term "modern Hereford" as uniformity of type for world markets. There was no difference in the conformation of a bull, or a female, required for any country. He had heard it said that the Argentine needed excessively short legs. That was not true. The Argentine needed size of a correct balance, meaning substance, leg length, and carcass conformation of a balanced volume. The butcher's block, said Mr. Schofield, was the deciding factor of a modern uniform type, and in the Argentine breeders had been taught by freight rates and the freezing plants the tremendous importance of uniform type. Uniformity of type was the key to success and prepotency was the key to uniformity in the modern Hereford.

Mr. Jack Turner thought the breeders in America and elsewhere were looking at Herefords with pretty much the same eyes. Ideas the world over seemed to be very much alike on questions of size and body conformation.

"In promoting the breed," said Mr. Turner, "we have got to do it not by trying to tear down someone else but by trying to improve our product, and if we do not make Herefords the cattle we think they are, and claim they are, we are not going to last. Let us keep the Herefords where they are most profitable—going down the middle of the road—and not be swayed by any fancies or fads."

Mr. Clifford (South Australia) congratulated the English Hereford breeders on the wonderful stock they are producing. He expressed pleasure that Polled Herefords were not bred in England.

Mr. Stafford Weston was pleased to hear from overseas breeders that they more or less agreed with English breeders on the type they wanted to breed.

Mr. Lindsay Field told the conference that there were 10,000,000 cattle in Australia, 3,000,000 of which were Herefords. Polled Herefords, he said, were making great strides in the Dominion and breeders of horned cattle throughout the world had to put their best foot forward to combat them. Mr. Field mentioned the heavy expense of importing cattle and suggested that the Hereford Herd Book Society should help overseas breeders in the way of cattle boxes. Mr. A. E. Baldwin, M. P. for the Leominster Division, assured Mr. Field that despite many obstacles and difficulties, the Livestock Export Group were doing all they could to ease the position.

It was at this stage of the proceedings that Mr. Lindsay Field brought forward the suggestion that the delegates should consider the formation of a world Hereford organization which could meet in turn in each member country and deal with international problems affecting the breed.

All delegates spoke in favor of such a plan, and it was decided to form a sub-committee to consider details and report to the final session of the conference.

The conference then came to what the chairman described as "the vexed question of registration." The subcommittee on this, he said, had had to meet twice, but they had succeeded in overcoming difficulties which had arisen at the last moment.

The resolution, proposed by Capt. R. S. de Q. Quincey and seconded by Mr. Jack Turner and carried unanimously, was as follows: "This committee recommends that the pedigree records of Hereford cattle registered previous to 12th July, 1951, in Great Britain, U. S. A., Canada, Argentina, Australia, Uruguay (New Zealand and South Africa, subject to investigation of their records) be accepted as between the countries aforementioned. The matter of the inclusion in their individual herd books of Polled Hereford cattle pedigrees be left to the discretion of the breed society, or association of each of these countries."

Delegates representing the various Hereford societies from around the world and their respective countries were: Argentina: Mr. James Schofield, of Chapel House, Appleby, Westmorland, Member of the Asociación Argentina de Criadores de Hereford; Mr. W. R. Schiele, of Los Molles, Buenos Aires, Member of the Asociación Argentina Criadores de Hereford; Australia: Mr. H. A Lindsay Field, of Eyton-on-Yarra, Healesville, Victoria, Member of the Australian Hereford Society. Canada: Mr. W. J. Edgar, of Innisfail, Alberta, President of the Canadian Hereford Association; D. A. Andrew of Calgary, Alberta, Secretary of the Canadian Hereford Association; Great Britain: Capt. R. S. de Q. Quincey, The Vern, Marden, Hereford, Member of the Council of the Hereford Herd Book Society; Mr. J. H. Everall, the Day House, Shrewsbury, Salop, Member of the Council of the Hereford Herd Book Society.

Ireland: Capt. S. F. Purdon, of Lisnabin, Co. Westmeath, President of Irish Hereford Breeders' Association; Lord Powerscourt, of Powerscourt, Enniskerry, Co. Wicklow, Member of the Irish Hereford Breeders' Association; New Zealand: Mr. R. H. Mead, of Wakefield, Nelson, South Island, Member of the New Zealand Hereford Cattle Breeders' Association; South Africa: Mr. A. Millar, O.B.E., Member of the Hereford Breeders' Society of South Africa; Mr. J. C. Conolly, of Figtree, Southern Rhodesia, Member of the Hereford Breeders' Society of South Africa.

Uruguay: Mr. Reginald Calvert-Booth, of Estancia Los Cerros de San Juan, Colonia, Member of Sociadad Criadores de Hereford, del Uruguay; U.S.A.: Dr. E. L. Scott, of Phoenix, Arizona, President of the American Hereford Association; Mr. J. Turner, of Kansas City, Missouri, Secretary of the American Hereford Association.

Entry of the APHA

Although the World Hereford Council had met regularly at four-year intervals since 1951, the American Polled Hereford Association was not invited to be a member. In spite of its annual recordings of more than 100,000 head of cattle, it had not been invited as an observer to the group. It had been established by resolution that no new members of the world group could be added to the council without the unanimous vote of every member association. By this means the American Hereford Association prevented an invitation from being extended to the American Polled Hereford Association. It is obvious that, as long as the American Hereford Association remained the sole member in the World Hereford Council from the United States, it had the exclusive privilege of certifying all Polled Herefords for export. Thus the AHA maintained a virtual club over the heads of Polled Hereford breeders, forcing them to register their cattle in the American Hereford Association in order to be eligible for world trade.

Several meetings were held by Polled Hereford breeders, including the board of directors, and many contacts were made with other members of the World Hereford Council to try to get the Polled Hereford story told and open the way for world recognition of Polled Hereford breeders. It is difficult for the relative newcomer on the Polled Hereford scene to realize the tremendous odds that Polled Hereford breeders found it necessary to battle over the years. It was common in that day for the powerful American Hereford Association to make sport of Polled Hereford breeders. It had the capability through power, prestige, and money to intimidate—to, figuratively speaking, suspend the Polled Hereford segment in the air and watch it slowly turn. It was futile for the Polled Hereford segment to think that it could meet its adversary head on. There were only two alternatives: one was to grow and wait to become superior in sheer numbers, and the other was to appeal to public sympathy. In some way the breed had to get its story told and present an honest appeal to world public opinion to free it from the burden and servitude.

One of the bizarre stories in the history of Polled Herefords is the happenings leading to the breed association's acceptance as a member of the World Hereford Council. The maneuvering on the part of its sister organization, the AHA, which blocked the entrance into the World Hereford Council is a story of intimidation and intrigue. Although the World Council is not a formidable organization and is more fraternal and promotional than legislative, there is one extremely important aspect about membership. It is necessary for mutual exchange of records in the case of exportations and trade in the world market. As long as the APHA could be denied membership, it was totally dependent upon the American Hereford Association to verify records on all cattle exported. This meant almost total reliance upon the AHA to record and issue its "official" certificate which would be recognized by the world group or a continuence by the breeders of the burdensome practice of double registry.

The AHA had a convenient interpretation of its rules of registry which required that all domestically produced animals must be continuously recorded in its records with no skips in pedigree. Oddly enough, they would accept the certified pedigree from sister associations in other countries but refused equal recognition to the APHA that resided in the same city. Obviously this was a design to force registry of all Polled Herefords in the AHA at regular fees, while the lion's share of the income from Polled Hereford breeders would be devoted to promoting the horned segment.

An unusual rule was adopted at the beginning by the World Hereford Council, which was operating without charter or by-laws at that time. After the initial meeting in England in 1951 it was agreed that the admission of any new members would require the unanimous vote of the members. One negative vote was all that was necessary to block a membership. The American Hereford Association was successful in preventing the APHA's application for membership from being considered. Many questions were raised on the part of other council members at the Third World Council meeting, held in Kansas City, Missouri, in 1960. Other members wondered why the APHA was not a participant. The minutes of that meeting indicate the unusual circumstance and the awkward sensitivity on the part of the foreign council members, who felt it strange that this sister organization, which had recorded more Hereford cattle annually than the rest of the world members combined, was not even invited as an observer or guest to the meeting being held in the city where it was located.

Finally in 1963 the joint contract between the two American groups was initiated. This contract provided for a joint certificate to be issued for Polled Herefords bearing the signatures of both United States association secretaries. This was in effect an endorsement of the APHA's records by the AHA, which removed the AHA's only legitimate reason for maintaining the adamant stand blocking the APHA's entry into the World Council. There were, however, roadblocks and hurdles still to be removed. After initiating a move to gain an invitation through the Canadian Hereford Association to attend the Fourth Conference in Dublin, Ireland, in 1964, application was made to the World Council headquarters in Hereford, England. The application was made to J. A. "Tony" Morrison, secretary-general, and copies were sent to all other council members. An attempt was made to gain the support of the American Hereford Association by asking the secretary, Paul Swaffar, to write a letter recommending acceptance of the APHA membership, but the request was denied. This was strong evidence that problems were still ahead at the conference table.

A large delegation of American Polled Hereford breeders was encouraged to make the trip to Dublin in June, 1964, to assist in a public-relations effort. About twenty-five breeders and their wives attended. At a breakfast meeting with Swaffar; Bill House, president of the AHA; Ben Smith, a Polled Hereford breeder and director of the AHA Board; Bob Swearingen, Sr., president of the APHA; R. E. "Pat" Connolly, a Polled Hereford breeder; and I, as executive secretary of the APHA, the AHA plan began to unfold.

On the morning of the opening session we were presented a written statement. Its content, if accepted, would have made the APHA the only member of the World Council with a conditional membership. It stated:

"Whereas, the AHA and the APHA have an agreement for the joint recording of Polled Herefords, and Whereas, the AHA is responsible for keeping all of the records of Herefords in the United States, Be it resolved that the APHA be accepted as a member of the World Hereford Council and that membership be in force so long as the contract between the AHA and the APHA is in effect. Be it further resolved that if the contract between the AHA and the APHA be discontined this membership should be reviewed."

The Polled Hereford group was appalled. The answer to the

statement was that it was nothing but an offer of second class membership, leaving our association as a pawn in the hands of the AHA.

I recall Swaffar's quick retort: "Well, Orville, we must protect ourselves in case the old crowd gets back in," meaning Chittenden, former secretary of the APHA, and the group of breeders who had resisted so strongly any tie-in with the AHA.

I said, "I know, Paul, but we have no assurance that you will be here forever, either." The meeting ended without agreement or any knowledge of how the AHA would react to our application when it appeared on the agenda.

There was no mistaking the fact that the AHA and Swaffar had firm control and great influence. That was obvious when, after submitting our statement and application, we were excused from the conference room by Lord Powerscourt, the chairman, while our application was being discussed.

Our membership was accepted. Swaffar was the first to emerge from the conference room and congratulate us. I later asked for a copy of the statement accepting our membership for verification from a recording secretary, and none was available. She typed on a piece of paper a statement that simply said, "Be it resolved that the American Polled Hereford Association be accepted as a member of the World Hereford Council." It was not until months later that we realized that the AHA had sown seeds of future problems.

The Summit

One of the most dramatic chapters in the history of Polled Herefords was written at the Fifth World Hereford Conference, held in Sydney, Australia, in April, 1968. To appreciate fully the importance of this drama and its impact on the breed, one must go back to 1963, when it all began. The five-year contract designed to relieve Polled Hereford breeders once and for all of the burden of double registry began on January 1, 1963. It was heralded as the beginning of a new relationship, an era when the breed would finally, after sixty-two years of divisiveness, be united. This period initially promised to be one of unity, harmony, and tranquillity. It could have been so—under one condition—if the Polled Hereford breeder would satisfy himself with being second. This would mean second place in a class with only two participants. The American Hereford Association insisted upon playing the role of "parent association": (1) owning the registry records, (2) determining eligibility for entry, (3) assessing fees, (4) declaring the rules for all open shows, and (5) selecting the judges—in fact, calling all the shots for the Hereford breed in America without any polled representation on its board.

The most serious and far-reaching effect of these conditions was that the horned breeder, with his traditional concepts, would establish type and performance standards, if any, for both segments of the breed. For polled breeders to succumb to this dominant influence would be to consign the progress of the breed to the whims of a tradition-bound group that had already declared its disdain for modern, progressive tools of breed improvement. The changes in breed philosophy so badly needed to encourage greater size and growthiness and move the breed away from the small, compact, inefficient Hereford that was the forerunner of genetic problems would never come about.

The decision was made to terminate the joint-certificate contract by the APHA in early 1967. Notice was given to the AHA to terminate, effective January 1, 1968. Battle lines were drawn, and the AHA began plans to have the APHA's World Hereford Conference membership nullified. If successful, this would result in virtually destroying the APHA as a recognized registry association by other world members. The AHA could then declare itself the authorized representative of all Herefords, polled and horned, in America. In that event the other members of the World Hereford Conference would no longer recognize the records of the APHA. Animals recorded only in the American Polled Hereford Association would not be eligible for export and entry in the records of other member countries. This would mean that after years of negotiation, hard work, and planning to eliminate double registration, double expense, and compounded paper work, and after having a five-year period of relief, Polled Hereford breeders would again be forced to resume the costly burden of double registration.

The Secretary-General of the World Hereford Council, Tony Morrison, had been kept informed as negotiations progressed with the AHA. He was aware of the joint-registry procedure and of the fact that in case of termination of the contract the authenticity of the records need not be affected. He was also aware that in case the APHA resumed its registry system it would constitute only a change in the place where the records were kept.

If this battle had been lost by the APHA, it would have been tragic not only for Polled Herefords in America but for the breed worldwide. There were two possible alternatives if the AHA succeeded in nullifying the APHA's membership in the World Hereford Council:

1. The Polled Hereford breeders of America could reconcile themselves to a future of horned dominance, and growth of this fastest-growing segment would be impaired. It would also have the net effect of retarding the growth of the total Hereford breed.

2. Polled Hereford breeders in America would probably separate themselves from the horned breeders completely without World Hereford Council recognition and ultimately develop a new breed and a new World Polled Hereford Council.

The latter course, obviously, would lead to strife and eventually a split in all other world member organizations. Many of them are too small to exist as complete economical units at present and would find it even more difficult if divided.

In fairness it should be pointed out that the AHA had a vital economic interest at stake in the question of Agenda Item No. 4 (quoted below). The prospect of losing approximately a half-million dollars in income a year from fees being paid by Polled Hereford breeders was not at all encouraging. Those funds, which had been in the past devoted to the promotion of horned Herefords would be sorely missed. Agenda Item No. 4 had an unprecedented impact on the breed worldwide. The outcome had most serious implications for Polled Herefords in America. It finally put to rest the question "What shall we do with the polls?" Since the world conference proved itself in handling this critical question, Agenda Item No. 4, it was a turning point in the history of this world body and resulted in strengthening and stabilizing the organization and setting the new direction for the future. This is the text of Item No. 4:

"To report termination of the agreement between the American Polled Hereford Association and the American Hereford Association, for joint processing of all Polled Hereford registrations and transfers, as from 1st January, 1968. (Item 4, World Hereford Conference 1964.) It was resolved: That the American Polled Hereford Association be admitted to membership in the World Hereford Conference Group so long as its records are acceptable to all members of the World Conference Group."

Because this Agenda Item was of the utmost importance to the future of Polled Herefords, excerpts from the transcript of the proceedings are presented here.

The Transcript

CHAIRMAN (Mr. R. S. Wilson, O. B. E., Treasurer, Australian Hereford Society): Before proceeding with this item I have to report to you that the American Hereford Association and the American Polled Hereford Association, have both submitted proposals to me concerning this resolution, and from those proposals we have worked out a summary of what we think would be desirable as far as this resolution is concerned. What we have drafted here is:

"The American Hereford Association wishes it to be noted for the record that reference to their Association as a Horned Hereford Association is incorrect, and that Dr. Berry is mistakenly referred to as an advisor to the Horned Hereford Association. They wish it to be clarified that the American Hereford Association serves, in fact, the whole Hereford breed, and feel that, while it is clearly too late to correct this in the program, it should be recorded at this juncture."

On the administrative side, we apologize for this error to the American Hereford Association, and, if the American Hereford Association is in agreement, I would request the Secretary-General to note this for the record.

Would the American Hereford Association be agreeable?

MR. DUEMELAND (Mr. Lorin Duemeland, American Hereford Association): Mr. Chairman, we are pleased that you have noted this on your records, and that it will be changed. We are very proud of our services to both the horned Herefords and Herefords without horns, and that is why we have asked for that change. Thank you.

CHAIRMAN: Thank you. Would you please note that, Mr. Secretary. As far as the American Polled Hereford Association is concerned, we have drafted this:

"That the American Polled Hereford Association wish me to draw attention to their position in that as full members, the World Hereford Group should recognize them as the official organization registering Polled Hereford cattle in the United States of America."

However, I do not feel that this Council is in a position to interfere in what appears to me to be a domestic situation, since we recognize both Associations as representing Hereford cattle, either polled or horned.

Is that acceptable to the American Polled Hereford Association?

MR. SWEET (Executive Secretary of the American Polled Hereford Association): Mr. Chairman, Distinguished Delegates; it was our opinion that it would be an asset to the other members of the World Hereford Conference if they had a clear understanding as to the areas of responsibility of the two Associations in America. We have the greatest respect for our co-workers, the American Hereford Association, but, as we study the history of the growth of Polled Herefords from their origin in 1901 we find that the American Polled Hereford Association came into being by necessity. By necessity because it was the only organization willing at that time to assume the responsibility of the recording of Polled Herefords and their ancestry.

I realize that time is of the essence in this Conference; but that we have thirty-three thousand breeders in America today that record one hundred and sixty-five thousand Polled Herefords a year, and they are vitally concerned with the deliberations here.

I think it is entirely a matter as to whether members of this Conference want to definitely identify the origin and the responsibility of the recording of these cattle or if they are willing to accept them from either.

Now, I think it is only fair that I make this statement at this time. The thirty-three thousand breeders of Polled Herefords in America met in December, 1967, and they resolved and identified the American Polled Hereford Association as their authorized representatives and as their official recording organization.

This has come about because of a series of evolutions in the last few years, when the American Hereford Association have identified themselves as promoters of horned Herefords. This has been done in promotional literature that had been widely distributed, I have one copy of it here if you would like to refer to it, and in the opinion of Polled Hereford breeders in America, the American Polled Hereford Association is their delegated representative and recording Association; and in this publication, which was published and distributed in January, 1967, Question 22:

"Why do true Herefords have horns?"

And the answer:

"That is like asking why does a 'he-man' have whiskers. The horns on gentle Herefords are nature's trade-marks of masculine vigor."

And it says:

"While the mysteries of genetics are not yet fully understood it is a truism of breeding hybrids and mutations that almost invariably when you breed for one trait you lose other profitable characteristics. That is why the horned Hereford does not suffer from genetic weaknesses common to many other outcross mutations."

I want to mention that there are reasons why the Polled Hereford breeders of America recognize the American Polled Hereford Association, and why this great block of breeders who are now recording approximately twenty percent of all the Herefords in the world, identify themselves and are determined that they shall be first, Hereford breeders, and second, promoters, breeders and improvers of Polled Herefords. At this time I would like to hand the microphone to Mr. Lewis, our President, who has also been President of the American Hereford Association, and also served as a delegate of that Association in 1960; he is well known throughout the world for his liberal and genuine, honest and basically unbiased views on the entire Hereford industry.

MR. LEWIS (Walter Lewis, Pres. of the American Polled Hereford Association, Larned, Kans.): Mr. Chairman, I come here with great admiration for this organization we have here today. It is through the workings of this great organization that we can make this breed of ours of Hereford cattle even greater.

It is time we have two segments within the breed, and I do not think it should concern any one of us as to who is sitting on his right or his left, as to whether he likes to breed Herefords with horns or without horns. I happen to be one of those breeding them without horns. I think I can say that the other members from our country can vouch that I have been a Hereford man in belief, but I preferred the more modern kind without horns, because they do have a real domestic value. It is much easier and more scientific to breed these horns off naturally than to go to the expense of doing it manually. I was born in this business, in the Polled Hereford business, and through the years we have had this problem within our country, where the Polled Herefords were good enough to stand the test, and I am willing to have you check the records that Polled Herefords today stand on their feet wherever they are shown in our country—so I do not think we should worry about whether they have horns or not. But, when we have a country like

Australia in size, and have so many people involved, and so many cattle recorded, there is plenty of room for both organizations. We have to worry more about other breeds than our Hereford breed, I think we have larger issues to consider today than a domestic issue in America, so let us get together and let each group take care of their own problems, work together and make Herefords still a greater beef breed which others are looking towards with admiration.

MR. DUEMELAND: Mr. Chairman, I would like to indulge on your time for just a moment because so many of you were warm friends with our Secretary, Paul Swaffar; I would like to bring you his greetings, and to say that we have with us today our new Secretary, Dr. W. Berry, better known as Dub, who is a man whom we felt was the only man available in the United States who could in any way fill the shoes of Mr. Swaffar. We are very pleased that Mr. Swaffar saw fit to stay with us and help train Dr. Berry in the Secretaryship of our American Hereford Association.

I would also like to take the time to tell you in some manner of our interpretation of the difficulty we find ourselves in today. As you know from the previous speakers, our five year joint contract was concluded on January 1st, or December 31st of this past year. It was a period of expansion of the Hereford breed, and, because the promotional job was divided during this period all breeders enjoyed the services of a separate field staff and a real period of growth.

During this period all eligible Herefords had their pedigrees brought up to date and put on magnetic tape to become part of our IBM computer record. The period served the Hereford well.

However, with proper notice the American Polled Hereford Association terminated the contract as of January 1st of this year, and the American Hereford Association as of January resumed even-handed service to all of the Hereford breeders, and, while some of them are small, it does include numerous polled breeders.

Those of us who sat around the discussion tables of the American Polled Hereford Association and the American Hereford Association, and that includes five of us who are here today, weighed this problem heavily, and we know full well that only time will show the results of the decision that was made at that time.

I believe, however, all of us agree that before anything else should come loyalty to Herefords. Anything that we do as segments, or branches, or departments should be done with an eye towards the total result, for in our country, the same as in many of your other countries, we are faced with new highly competitive breeds, and even the acceptance by the commercial breeder of un-registered sires.

Herefords are beginning their third century and, if you will permit my own personal observation, the first century was concerned pretty much with the development of the Hereford, the second with the stabilization, and the third will be concerned with the expansion, and much of this will depend on a free flow of stud stock between nations.

All this leaves us with what we call knotty problems, for, while our funds, so necessary for research and promotion, will be lessened, we promise that we will continue our work on dwarfism, pink eye and cancer, and our recent purchase of an IBM computer will allow us to expand our performance and carcass program, which we call total performance records. Article 3 of our Articles of Incorporation charges us with the responsibility of recording all registrations and transfers of all pure bred Herefords in the United States. Our Rules provide that, with the exception of records of member countries, we must register only calves that are born domestically, whose registered sire and dam appear in consecutive order on our records. When a calf goes unrecorded on our magnetic tape, the progeny cannot be entered, because we would then not have a continuous record.

We have consulted our attorneys, who, in their considered opinion, point out the inability to delegate this authority to any other group without direct supervision, and so it looks like the American Hereford Association, which records Herefords both with horns and without horns, is faced with a separate Association which records by its own system Polled Herefords.

In the back of the manual that has been presented to us today it lists the Rules of the World Hereford Council to be considered as Item 20. We find at the top of page three a list of three categories under Eligibility for Membership:

(a) An Association or Society being the Registration Authority of a country;

(b) An Association or Society in any country which, although not the Registration Authority, is the sole representative of Hereford cattle in that country; or

(c) An Association or Society in a country whose records are maintained by an existing member country.

At the bottom of page six there is (b) under Pedigree Records. This paragraph states:

(b) Registrations or transfers of cattle recorded by members of the World Hereford Council must be honored and accepted in the registrations and Herd Books of all other participating Associations and Societies.

You can see that we in the United States will be faced with a problem. Now, if you would refer back to page three, the second paragraph, it says:

"Note: An Association or Society representing a minority or special interest of Hereford breeders will not normally be considered for membership. Where the Registration Authority and the Hereford Association in any one country are the same no other group will be eligible for membership, but may be admitted to Associate membership."

I would like to conclude this with a very short story that I saw on television recently—a Negro comedian was doing a dance, and when he finished the dance he said, "I do not have the solution, I am part of the problem," and in our country that is a problem. I thank you, Mr. Chairman, for the time given by the Council for us to make this statement.

MR. SWEET: Just one brief statement. I want, first, to apologize for our having to discuss this issue here, and I indicate our appreciation for your indulgence while we listen to this problem, it has been with us since 1901, so I guess you might say we are the problem. We have been a problem for many years, and it seems the Conference has had some difficulty in years past in determining just what to do with us.

I am amazed at the ease with which three of our members were made members today—Mexico, Portugal and Spain. I think within thirty minutes time we had checked their credentials and they received a unanimous vote. Unfortunately, it took the American Polled Hereford Association thirteen years to be even invited as observers.

I think we can dispense with this without any real difficulty and without anyone getting bruised up. In the first place, it is our interpretation that the World Conference does not deal in domestic issues, nor is it mandatory that records be reciprocal on a domestic basis.

Australia has, in my opinion, one of the finest working relationships at this time. They have two Societies with responsibilities, and do not have reciprocal acceptance of records. It is not our feeling that everyone should be as we are. We have different conditions. Australia and the United States are both large countries with great numbers of members. We have thirty-three thousand. We have over one thousand Polled Hereford events each year, four hundred of them are Production Sales; so, the responsibility in the United States must be divided up in some way, whether north and south, east and west, polled and horned.

This does not mean that we have to have difficulty, that we must be enemies. I think this one point is worthy of your consideration.

SECRETARY-GENERAL (Mr. J. A. Morrison, Hereford Herd Book Society, Hereford, England): For the record, Mr. Sweet, the three new members were not elected in thirty minutes of looking at their records here. These were examined over the last few years very carefully.

MR. SWEET: I am sorry if I implied, and I certainly did not mean it as being disparaging, because I am as aware as you of the work that you have done in checking their credentials beforehand, but the real problem is defining the areas of responsibility. The only records in the world that trace back to the original Polled Hereford are with the American Polled Hereford Association.

Mr. Swaffar states in the 1956 Conference—at page seventy-two:

"You should know that our Association could not give you a fifth generation pedigree on an animal that may have ancestors prior to 1952, and guarantee that it has no polled blood in it, unless you happen to know it, or unless you had a certificate from the Polled Association."

It came into being, our Association, because of a necessity to keep Polled Hereford records, and these records have been intact. Also, in that meeting Mr. Swaffar implied that there may be a problem with our records, but he did not know exactly what it was, and, of course, the implication was that single standard animals were still being recorded. I hardly see how he could not have known at that time, in 1949, when a publicized statement was made and a resolution was passed by the American Polled Hereford Association, in which all of the records of single standard animals were sealed up, and all animals from that time forward traced to Volume XIII, or prior volumes.

Mr. Duemeland referred to the expired contract. We exerted every effort, and with due credit to the accredited delegates of the

American Hereford Association—they were conscientious in their efforts, but I think one statement is very significant. The proposal written by the American Hereford Association said this:

"It is desirable to extend the joint certificate agreement for another five year period, provided the financial position of the American Hereford Association can remain equal to that provided under the present agreement."

Here again we are dealing with domestic problems. I would like to re-emphasize that the reciprocity of records on a domestic basis insofar as we can determine, Mr. Secretary, was not intended by the World Hereford Conference Group. Its control extended only between members of the World Conference—is that not true?

SECRETARY-GENERAL: Could you put it more clearly, please?

MR. SWEET: Is it not true that the intention of the World Society was to arrange for reciprocal acceptance of records between member countries on an international basis, and the arrangement made at the 1956 Conference was not intended to be a control or to arrange for a reciprocal trade of records within a country?

SECRETARY-GENERAL: This particular point that Mr. Sweet makes, Mr. Chairman, was not defined as such, so he could only be referring to an intention.

MR. MATTHEWS (Mr. W. I. Matthews, New Zealand): It does appear to me to be a domestic issue, and I would like to offer a resolution for the consideration of the Conference, in an endeavor to solve this question.

As a preface I would like to say that the American Polled Hereford Association change in procedure has in no way affected its eligibility requirements for registration and transfer which are embodied in the by-laws of the Association, and the Secretary-General of the World Hereford Council has advised the American Polled Hereford Association that the integrity and accuracy of their records—and this is a quote—do not require re-verification of recording procedures.

Therefore, I offer this resolution for consideration of the Conference:

"That the matter of procedures for registration is essentially a domestic matter, but that the World Hereford Council finds no fault in the procedures adopted by the American Polled Hereford Association."

CHAIRMAN: Thank You, Mr. Matthews. I will accept that resolution. Could I have a seconder?

MR. HAWKINS (Mr. Keith Hawkins, Australian Poll Hereford Society, Brisbane, Australia): I second. I have listened intently to all the discussions from our American friends. I can see that they have a problem. I also see that it is a domestic problem, and not one to be resolved here at this Conference. We, in Australia, have been very fortunate to have two separate Societies with two Secretaries who are bosom friends. We have two Societies which pull together to promote white faced cattle, Hereford cattle, in this country.

I think it is very important, as Mr. Duemeland mentioned earlier, that we try to improve and hold our position with Hereford cattle against these modern breeds, which are improving all the time, and challenging this breed from the supremacy it holds throughout the world.

I would like to second Mr. Matthews' motion that the matter of procedure for registration is essentially a domestic matter, but that the World Hereford Council finds no fault in the procedures adopted by the American Polled Hereford Association.

SR. BACA (Sr. Julio Baca, Asociación Hereford Mexicana A.C.): Although the differences between the American Hereford Association and the American Polled Hereford Association are domestic problems of the United States, Mexican law requires that all the imported registered Hereford cattle be registered with the American Hereford Association, and I think that it is the responsibility of this Conference that this issue should be solved.

MR. COULTES (Mr. J. R. Coultes, Canadian Hereford Association): We, in Canada, have viewed this problem in the United States as a domestic problem, but today we are not sure that this is so. We are led to believe by our Registrar in Ottawa and others that, since it has been brought to our notice this Item 4, it is not now a domestic problem.

As we understand it in Canada, when an American Polled Hereford Association animal is imported from the United States and registered only by the American Polled Hereford Association as we call the single standard registration, then it goes through our books, and in the period of the time the progeny from this importation is sold back to the United States to a member of the American Hereford Association, this animal would not be eligible for registration in the United States.

As we have considerable trade back and forth across this line, this would jeopardize the Hereford industry between Canada and the United States. Just as recently as a week or two before we came down here, at the Calgary Bull Sale, there were sold somewhere in the neighborhood of eight hundred bulls, eighty of which went back to the United States. If these animals had poll blood of single registration by the American Polled Hereford Association they would not have been eligible, as I understand Item 4 of the Charter of the United States, and, therefore, we in Canada do not think that it is a domestic problem. It has domestic features, but it involves Canada to the north and Mexico to the south.

SR. BACA: Mexican law does not allow animals to be imported from anywhere in the world but Canada and the United States. The Hereford breed in Mexico is the largest in the country, commercial cattle that is. Our registered cattle have to be imported from those two countries, and Mexico needs to import cattle, and also to assure our members of accurate and continuous records. The problem is not, therefore, a domestic one, but a problem between Canada, the United States and Mexico.

CHAIRMAN: I would like to ask a question of Mr. Coultes. In the instance that he put forward, would not those animals that are re-exported, or the progeny of those animals that are exported to the United States, be eligible for registration with the American Polled Hereford Association?

MR. COULTES: Mr. Chairman, I understand that they would be eligible for registration with the American Polled Hereford Association, but not with the American Hereford Association. That is my understanding and that is Canada's understanding.

We have it in writing, but let us go further. We have, recently had importations from the United States with single registrations going back to the time when they had single registrations seven or eight years ago, which our Registrar refused to accept. If they are not acceptable in our records so that we can re-export their progeny our Registrar will not accept them. We have asked the authorities over there, and we have it in writing. It is just as simple as that. If I import single registration animals I cannot export the progeny back because, as Mr. Duemeland pointed out, there would not be continuous records with the American Hereford Association.

MR. SWEET: We are dealing, in the first place, with some terminology which is ambiguous. We are talking about single standard animals. There is not an animal registered in the Hereford world today which is not a single standard animal. Your Canadian Association animals are single standard, they are registered only once. The American Hereford Association animals are single standard, registered only once. As long as these animals trace back to Volume XIII, or prior volumes, then they should be eligible in any Book of any of the members of this Council. If that is not so then those members that refuse them, logically, would be jeopardizing their own membership.

So far as Mexico is concerned, when I voted for the Delegates' membership, I was not aware of the fact that they had such restrictive Rules. I think this is something that we should take a long look at.

Insofar as the American Hereford Association Rules are concerned, I have made a detailed study of their Charter as to Rules of registration. Their Charter is right here, and it says nothing whatsoever about an animal being imported having to trace to animals that were previously recorded in the American Hereford Association. It does refer in one instance that animals. . . . "From time to time Rules will be written and distributed to members," and I have a copy of the most recent Rules that have been circulated to members, and again, there is no objection, no blockage, for animals coming from Canada to the United States that have blood or background of the American Polled Hereford Association, so long as they trace to Volume XIII, or prior volumes. Mr. Swaffar wrote in June to the Canadian Hereford Association and indicated to them that if they accepted our records it would also jeopardize theirs, that they might not accept the Canadian records.

I think we might as well be honest and keep the integrity and the logical sense in this Conference that started in 1951. We are dealing with an economic problem here. The American Hereford Association are very good friends and co-workers, we would like to establish a relationship with them, we have offered to co-sponsor any and all events in the United States, and assume our share of the financial responsibility; we offered to continue our registration program with them and pay a pro rata share of the expenses. We are dealing with an organization that wants to maintain its income; I can understand this.

But, you are also dealing with a segment that is determined, and has been since 1901, to survive, and to take care of their business. We take great pride in being a part of this organization, and when we submitted our records and application in 1964, we accepted membership in good faith. It was indicated to us that it was a full membership, and no one has brought anything forward to us to indicate that our membership be jeopardized. We submit our records to you today, and petition you to look at this issue in all fairness, and to keep one thing in mind—that we are not now discussing whether or not the American Polled Hereford Association should continue recording or the American Hereford Association should continue to record Polled Herefords; what we are dealing with is the integrity, the intent and motivation behind this Conference—is it to be used as a tool for the few, or is it to be used by all of us as a great tool to continue to improve and promote Herefords?

MR. CHAIRMAN: I am afraid at this stage I find that I have been a little lenient, and I must remind you that from now on I will only allow one person to speak once and once only to any resolution, except the mover who would, naturally, have the right of reply. Is there any further debate?

MR. HUMPHREYS (Mr. F. E. Humphreys, New Zealand Hereford Breeders' Association): I feel, Mr. Chairman, that this is a problem that has come about through the growth of our breed in the country that has more Herefords than any other country in the world, and I think we should let them solve their own problems. I think probably the numbers of Herefords are so great that their records would be much better kept separately in their two different organizations, and I feel if we pass this resolution we are doing something for the good of the breed.

CHAIRMAN: Any further speakers?

SR. IRAOLA (Sr. Pereyra Iraola, Asociación Argentina Criadores de Hereford, Buenos Aires, Argentina): Argentine law requires that Certificates can only be given by the American Association, it is their property, and they are the only ones that can give it. If we do not have the Certificates, we would have to close our doors in the Argentine. I believe that the problem we are discussing is not a problem affecting only the United States, but all of us, due to this division and separation of pedigrees.

MR. HANSON (Mr. J. A. Hanson, Australian Poll Hereford Breeders Society, Brisbane, Australia): I very much regret to hear these discussions. I am sure it is only a domestic problem, and I notice from the Delegate from Canada's remarks—the purity of these cattle is not doubted and they are all traced to Volume XIII. The keeping of their records is not doubted. There must be something wrong with our Conference Rules when Canada cannot alter her Rules to fall in with all this, because the cattle are pure and the records kept right. I feel if Canada cannot register these American Polled Hereford Association cattle with the American Hereford Association, why not change their Rules so they can?

MR. ERICKSON (Mr. C. B. Erickson, Canadian Hereford Association): There is no question of Canada not being able to register their American Polled Certificates, it is just a question of being able to re-sell the animals wherever they wish in the United States.

MR. LEWIS: I would like to answer the question, because I live across the border and we do sell bulls and females in Canada. I can assure you I have no trouble in bringing them back into the United States. You just give us a try. Those cattle will be accepted in the American Polled Hereford Association just the same as if they were living south of the border and I dare you to point me out a breeder who is breeding horned cattle that is going to buy cattle back from Canada back into a horned herd. When a man changes to breed polled cattle he will stay with the organization he has switched to, so I see no problem whatsoever.

CHAIRMAN: I think you have had sufficient debate on this.

MR. HOUSE: (Mr. Bill House, American Hereford Association): I do not want to lose my right to speak to the subject, but I question the Rules and would like it established by the Chair

what we are trying to vote on. If it is a determination of whether
we accept another Association or not, that is one matter; if it is
simply a resolution saying it is a private matter, this is another;
and, before there is any voting, I would like a re-statement on
what we are voting, and what it determines, because I do not want
to lose my right to discuss the matter, if it is whether or not to
accept a separate organization from the United States into this
Conference.

CHAIRMAN: We are not voting on any matter concerning any
particular organization. Both Associations, as I understand it,
within America, are full members of this World Hereford
Council. The resolution virtually says this is a private matter and
that is all we are voting on.

SECRETARY-GENERAL: For clarification—I think the subject has
gotten a little out of hand—this Item 4 was merely to report the
termination of the Agreement between the two Associations in
the United States, and that, as far as this Council was concerned,
it was merely noting that there had been a change in the location
and management of the records concerned. That was our only
involvement in this, but owing to the fact that both the American
Associations presented a statement to the Chairman before the
meeting, I think we are now discussing the two statements given
to the Chairman.

We do not appear to have reached the Report, which simply
notes there has been a change in the location and management of
the records concerned. A resolution appears to be emerging from
the statements made by both organizations to the Chairman.

CHAIRMAN: This is correct, and when we finish dealing with
this resolution at the present time, I would then ask for a
resolution noting the termination of the Agreement, as set out in
your papers under Agenda Item 4.

MR. MATTHEWS: We have a resolution before us.

CHAIRMAN: That is correct—that is the one we are debating.

MR. HOUSE: Would the Chairman please read the resolution again?

MR. MATTHEWS: Could I read the resolution again? There does appear to be some confusion as to the actual meaning of it. The resolution is:

"That the matter of procedure for registration is essentially a domestic matter, but that the World Hereford Council," and I think this is the important part of it, "finds no fault in the procedures adopted by the American Polled Hereford Association."

Just before I conclude I would like to remind the Conference of Item 5, the Charter of the World Hereford Council, and that is;

"To maintain a high standard of ethics in all promotional activities both at individual and Association levels."

I think this does concern some of the discussion we have listened to.

MR. HAWKINS: As seconder of the motion, could I speak?

CHAIRMAN: No, you spoke when you seconded it. Is there any further debate?

MR. HOUSE: I would like to take my turn this time, Mr. Chairman. We would like to make it clear that the original authorization and membership admittance was based on the fact that the records were in our hands and under our supervision as the American Hereford Association, and I thought perhaps at the time that, if we came down here, then there would be a clearly established question as to whether or not to admit a separate Association, which they are, in fact establishing in the United States at this time.

Even if the American Hereford Association chose to agree with them fully in everything they are doing in setting up a separate registration outside our supervision, we cannot treat them any differently than we do the members who record consistently with us, and the one Rule that has been established back in the 1800's was that the sire and dam have to be on our records, or we do not accept the progeny, and it does not make any difference who it is. This Rule is fixed, and we have followed it consistently, and there is no indication that it will be changed. Any domestic born animal that wants to come and be recorded in the American Hereford Association must have both the sire and dam recorded, and it must be a consistent recording all down the line.

There are two reasons for this—the first is that we can supervise our own records under our Charter, which was the original, and at the one time the exclusive right; secondly, we cannot trace parentage in seeking dwarfism possibilities unless we have this.

This poses a very special problem for us, because everything has to be fed onto our tape out of our own records, and we cannot furnish any information if it is not consecutive and in order on our own tapes.

If we are trying to approve a separate organization, I want it made plain to this group for the Conference, so that they will know what they are voting on, because, if we move in this direction and determine that you are going to accept a separate organization in the United States, you are actually, according to the publication which they sent to their own breeders, in answer to a question:

"Will the American Polled Hereford Association record horned offspring?"

They answered:

"Yes. Horned offspring will be eligible for registration only when one or the other parent is polled. They will be recorded and added to an appendix type record provided for horned cattle. A certificate will be issued the same as for polled cattle except the registration number will include an 'H' suffix."

In other words, whether we recognize all their problems, we want it made clear that this is supporting a separation, and that you are taking in two separate organizations within the United

States for membership.

It also violates one principle that we have, and forces us to turn around and say we cannot accept from them a domestic born animal, because we have an exclusive Certificate or Charter originally, and it violates our Rules.

It almost forces us to take a position, perhaps even, that we would not like to take, and the only thing we are quite anxious to do is to see that you folks understand what you are doing, make your positions clear, so that after the Conference is over you will not say that you did not realize what you were doing.

We are getting many applications from foreign countries to provide them with a Certificate concerning dwarfism. This again poses a problem, and we want it understood that we cannot furnish these certificates unless there is consecutive recording in the American Hereford Association, so that our records are clear and consistent.

MR. SWEET: Reference was made to the . . .

CHAIRMAN: You have had three turns on this subject, and I ruled before I would not let you speak again.

MR. SWEET: I would like to answer the question.

CHAIRMAN: I am sorry, I must stick to the rules of the debate as I see them.

MR. HAWKINS: A question to Mr. House—What happens to an animal imported from England? He has stated sire and dam must be on the record of the American Hereford Association.

MR. HOUSE: Our Rules apply only to domestic born animals. We do not try to reach outside.

CHAIRMAN: I propose now to give Mr. Matthews of New Zealand the right of reply.

MR. MATTHEWS: I do not wish to exercise the right of reply, Mr. Chairman. [On being put to the vote the motion was carried, nine votes for and six votes against.]

It was resolved:

"That the matter of procedures for registration is essentially a domestic matter, but that the World Hereford Council finds no fault in the procedures adopted by the American Polled Hereford Association."

CHAIRMAN: Will someone please move that we note the termination of the Agreement between the American Polled Hereford Association—that the American Polled Hereford Association has decided not to renew their joint Contract with the American Hereford Association?

MR. FIELD: I so move.

MR. HAWKINS: I second.

CHAIRMAN: This is noting that the agreement has been terminated.

SECRETARY-GENERAL: I would like to stress that at the last World Conference, the American Polled Hereford Association were made full members of this organization, provided their records were acceptable to all members of the World Group. The records we assume are basically the same since they were supported at that time by the American Hereford Association. This is merely a note for the record.

MR. HOUSE: I would like to draw the Secretary-General's attention to page 121 of the Conference proceedings in 1964, which reads:

"It was resolved that the American Polled Hereford Association be admitted to membership in the World Hereford Conference Group so long as its records are acceptable to all members of that Group."

And then you should refer back to the discussion previously which shows at that time the American Hereford Association was keeping records, so it is plain that the problem has arisen because the records and locations have changed from one Association to another.

CHAIRMAN: All we are doing is asking that this Conference

notes that this has happened. Are there any further speakers?

SECRETARY-GENERAL: In answer to Mr. House's point, and he raises an interesting question in that the American Polled Hereford Association were, nevertheless, accepted as full members, and as we sit around this table now, they are full members. We have noted that their records have moved from one place to another and are under different administration.

MR. HOUSE: On page 121, it says:

"So long as these records are acceptable to all members of the World Hereford Conference Group."

We have tried to make it plain that under our by-laws and our Charter and our Rules we cannot accept skip registrations into the American Hereford Association records, both for the Rules and because we could not furnish the kind of material we have been asked to on dwarfism in Herefords.

MR. SWEET: In answer to Mr. House, the American Polled Hereford Association has the most complete records on dwarfism as far as Polled Herefords are concerned. We do not issue a certified pedigree or a certificate for export unless it passes our pedigree check. On Mr. House's statements referring to membership, in the first place, if you refer to page 121, Mr. House was asked a question by Major Ponsonby:

"Can I ask Mr. House or Mr. Swaffar about this resolution; are the Polled Hereford Association in full agreement with it?"

And Mr. House's reply was:

"That is my understanding. We prepared this jointly, and to the best of my opinion we are in complete agreement about it. I would be happy to read it to you."

I would like to make this clear—that the resolution which was drawn at the time was not submitted to the American Polled Hereford Association, we did not help to draw that resolution. We saw one that was prepared the day before which indicated to us that there might be conditional membership attached. We said that we had been authorized by a Board of Directors to apply for and receive full membership, nothing more and nothing less. Before we were excused from the Conference Hall Mr. Hawkins, the delegate from Australia, asked us if we would be willing to accept membership on that basis, and we said, absolutely not.

As soon as the Conference was over, I went to the Secretary and asked for a copy of the resolution which was read today:

"That the American Polled Hereford Association be admitted as full members of the World Hereford Conference Group . . . " and that is the resolution that you see in the summary of the Conference proceedings.

CHAIRMAN: Thank you, Mr. Sweet. It was agreed:

"That the termination of the agreement between the American Polled Hereford Association and the American Hereford Association, for joint processing of all Polled Hereford Registrations and Transfers, as from 1st January, 1968, is noted."

(Author's note: The delegates representing the American Polled Hereford breeders assumed that the issue of APHA membership was closed after Agenda Item No. 4 was voted on and passed on Tuesday, April 2.

(The representatives of the AHA were obviously unhappy with the turn of events and were not willing to give up so easily. Thus the question was raised again on Wednesday, April 3.

(Not included in the official minutes of the conference was considerable discussion about the chairman's method of counting votes. Traditionally the conference had conducted its affairs by the "one man, one vote" sytem. This system had been questioned by Sr. Pereyra Iraola of Argentina, who felt that each country should have two votes even though it had only one representative present. He also indicated he would like to vote the two proxy votes for Uruguay, which had no representatives present.

(If this were allowed and Argentina and Mexico were allowed to vote proxy votes, it would have shifted the balance in favor of the AHA, and our acceptance as continued full members would have been nullified. With this in mind, the move was made to reopen the discussion on Agenda Item No. 4.

(It was imperative that reopening Agenda Item No. 4 be delayed in order to inform the chairman of the traditional method used by all previous Conferences and that this be explained to him by the most appropriate persons.

(The spontaneous decision by the APHA delegates to invoke the use of the "substitute motion" to stall any further discussion of the subject of APHA membership was a strategic move at the most critical moment. The move resulted in further delay of discussion on Agenda Item No. 4, and ultimately the chairman was informed of the traditional "one man, one vote" precedent.

(The conference was held over an additional day for a special session on Thursday, April 4, to reopen Agenda No. 4. The minutes continue and give an accurate account of that session. The Polled Hereford delegates came out victorious, as the following minutes reveal.)

[Item Number 4 on the Agenda was reverted to at a Working Session on Wednesday the 3rd April, 1968, after Item 19. Reverting to: Item No. 4. To report termination of the agreement between the American Polled Hereford Association and the American Hereford Association for joint processing of all Polled Hereford Registrations and Transfers, as from 1st January, 1968. (Item 4, World Hereford Conference 1964)—It was resolved: "That the American Polled Hereford Association be admitted to membership in the World Hereford Conference Group so long as its records are acceptable to all members of the World Conference Group."]

SR. BACA: I would like to comment on the Chairman's

apologies for Mexico at the beginning of the Session. I will accept such apology and make it a motion: "That a new vote count be taken of Item 4 in order to keep our records straight."

SR. IRAOLA: I second Sr. Baca's proposal.

CHAIRMAN: I have before me at the present time some notes which have been passed to me by the American Hereford Association.

I was trying to clean up the Agenda first before we got back to reconsideration of Agenda Item 4. I have an explanation in front of me now which I think would probably clean up Agenda Item 4 to everyone's satisfaction—I sincerely hope so. If it is your desire we will do it now.

MR. FIELD: Could I suggest that it be done after Item 20?

SR. BACA: I moved a motion and it was seconded by Sr. Iraola.

CHAIRMAN: Do you really want to discuss Agenda Item 4 again now, or are you prepared to deal with Agenda Item 20 first?

MR. BACA: Now we have started talking on it, I think we should go through with it.

MR. SWEET: I think it is perfectly in order and according to the International Parliamentary Rules of Order, and I would like to offer a substitute motion.

CHAIRMAN: Before I can accept any motion as far as Agenda Item 4 is concerned, I would have to have a motion of recommittal, and before we do that, can I allow the Secretary-General to read the notes that we have at the present time?

SECRETARY-GENERAL: Ten minutes before this Session started, I was handed by the American Hereford Association some notes here which include a resolution.

May I just say before reading it that I think in this particular situation we have all felt slightly uncomfortable and rather sad that there should be this difference of opinion, indeed a rift, between our two American Associations, representing the greatest concentration of registered pedigree Hereford cattle in the world, and I know we all wish this matter to be resolved before we leave the shores of this exhilarating country, in which we have all attempted to make, I am sure you will agree, a very positive contribution towards the problems now facing the world in the production of beef.

I have been passed this resolution, which appears to me—at any rate—to be a very sincere attempt to resolve the matter in order that we may leave this Conference united in our outlook and, therefore, better able to face the important problems at global level during the next four years.

As far as your Secretariat is, and must be, concerned, may I repeat:

1. That the American Polled Hereford Association are full members of this World Group, by resolution of the Fourth World Hereford Conference;

2. Their records have been accepted as being authentic and the method of maintenance as being satisfactory for reciprocal registration purposes;

3. The mechanical procedure of maintaining the records is unchanged;

4. The only change is in their location and management; and I would urge you for the sake of our breed to resolve this unhappy matter, since we are at an important crossroads in our history.

Here are the notes received from the American Hereford Association:

"Realizing that some confusion may have been created yesterday by Item No. 4 on the Agenda, which acknowledged the change in recording some Polled Herefords, it remains a necessity for each member country to take home a firm interpretation of the World Hereford Council's actions;

"(a) The United States domestic problem involved thousands of Polled Herefords now being registered by two separate Associations;

"(b) International trade involves only a limited number of these animals, but becomes highly significant, because several of our World Hereford Council members have specific Federal Laws, which prevent their breeders from dealing with two Associations within the United States.

"Therefore, we wish to put these thoughts into this motion, in an attempt to resolve this situation:

"Be it resolved;

" 'That this Conference, having duly noted the January 1st, 1964 change, permitting polled animals to be recorded in two separate Associations in the United States of America—at the breeder's option—and further taking note that sales within the United States of America are a domestic problem. We move, however, that all Herefords exported from the United States of America be officially registered also by the American Hereford Association to meet these foreign requirements.' "

That is the suggestion, which may, or may not, be acceptable, but it is, I believe a genuine attempt to resolve the problem.

MR. SWEET: I rise to a point of order. I am prepared to offer a substitute motion which, according to the rules, takes precedence over an original motion.

CHAIRMAN: Before I am prepared to accept any motion concerning Agenda Item 4, whether it be the one as read out by the Secretary General, or whether it be a substitute motion, as suggested by Mr. Sweet, I would be pleased if someone would move that Agenda Item 4 be recommitted.

MR. SWEET: I will move.

MR. COULTES: I will second . . .

MR. SWEET: I am not moving that Agenda Item 4 be re-submitted.

CHAIRMAN: I am sorry.

MR. SWEET: My substitute would be:

"That all other business be temporarily delayed until we deal with Item 20."

I would appreciate a seconder.

CHAIRMAN: Mr. Sweet has moved that all other business be temporarily suspended until we deal with Item 20. I did not understand what you meant by a substitute motion.

MR. HAWKINS: I second.

CHAIRMAN: You have heard the resolution as put up by Mr. Sweet and seconded by Mr. Hawkins. It is a motion of procedure.

It was resolved:

"That all other business be temporarily delayed until item 20 has been dealt with . . ."

[Item Number 4 on the Agenda was re-opened at a Working Session on Thursday the 4th April, 1968. Reverting to: Item No. 4. To report termination of the agreement between the American Polled Hereford Association and the American Hereford Association, for joint processing of all Polled Hereford registrations and transfers, as from 1st January, 1968. (Item 4, World Hereford Conference 1964)—It was resolved: "That the American Polled Hereford Association be admitted to membership in the World Hereford Conference Group so long as its records are acceptable to all members of the World Conference Group."]

CHAIRMAN: Gentlemen, I was waiting to make sure all were present, for the simple reason that, before we can re-open this discussion, I must have exactly the same personnel—as far as voting powers are concerned—in the room as there were on Tuesday afternoon. I feel that we are meeting this morning in some form of tension, which I think possibly is not quite as good as it might be for this organization, but I do appeal to you all that when we leave this room this morning we leave it in peace and harmony, and do not forget what we said last night, that the eyes of the world today are on this Fifth World Hereford Conference, and whatever decisions we arrive at this morning, when we walk out of that door I trust we will shut the door behind us. I am not very good at making appealing speeches, and that sort of thing, but there are a few things I must say first. One is that we are meeting in a different room, you have changed your positions, we have not got the simultaneous translation that we had in the other room and, so that we may have a complete record, when you speak will you please give your name and that of your country.

It has been suggested to me by some that they do not quite understand some of the procedure which I have adopted at this meeting. What I intend to do this morning—and which in my book of rules of procedure is the only way to handle it—is that, before this matter can be discussed, we have to realize that we have a resolution in our records which was moved by Mr. Matthews of New Zealand, and seconded by Mr. Hawkins of Australia, and before this can be discussed, I will ask shortly for a procedural motion that this resolution under Agenda Item 4 be re-committed.

Being a procedural motion, I will then have to put it to the vote, when, if that procedural motion is not carried, I will have no other way except to declare this meeting closed; if it is carried, we will then have that resolution again before the meeting, and I will accept any amendments to that resolution, unless the mover and seconder request permission from me to withdraw it. Is that clear to you all?

MR. HOUSE: Mr. Chairman, do I understand you that you will accept no other resolutions or no other motions concerning this matter at all if this one loses?

CHAIRMAN: That is correct.

MR. HOUSE: And that you will let the Conference be abandoned without any attempt at conciliation, or a definition of how we left this?

CHAIRMAN: Mr. House, I feel I have no other way of handling it correctly under meeting procedure, but I feel sure that it would be the desire of everyone that we clean this slate.

MR. HOUSE: Would it then, if this was handled in this way, be the privilege of this group to ask you to hear a motion that I might present that would clear up some of the problems that have been presented at this time?

CHAIRMAN: If it so happened that this re-committal was not carried, I still have "General Business" before this meeting.

MR. HOUSE: And you would accept a motion?

CHAIRMAN: I would accept a motion after it has been read to me—I would reserve the right to accept it or not.

MR. HOUSE: This would be fine, and what I am getting at is I do not want to go home and try to tell our folks that we simply were not heard, because that threatens the very foundation of the existence and our membership, then, of the Conference if we are cut off in that manner.

CHAIRMAN: If it had to come up under "General Business", so long as it was not contrary to the resolution we have on our slate at the present time, I would accept it; but if it is contrary to it, I would not be in a position to accept it.

MR. DUEMELAND: Would this also then include the motion that was put by the Secretary-General on our behalf on Tuesday, which was delayed so that the meeting could be completed through the Rules yesterday?

CHAIRMAN: If my memory is correct, what the Secretary-General read out was not put before the meeting. I did not have it before the meeting as a motion. I could not take it because Agenda Item 4 had not been re-opened.

MR. HOUSE: Was the motion that was made and seconded to correct the records on the counting, is it still in the business of the meeting?

MR. SWEET: The question that was brought up was in connection with the motion for re-counting the vote, and that is the motion which we replaced with a substitute motion, and which was nullified; so that motion, according to the procedure, we assume, depending on your records, was removed from the books.

CHAIRMAN: I did indicate to some delegates last night when they asked me, that I would give this decision this morning. I would hope that we would not arrive at a position where we have to vote this morning, but if we do arrive at that position where we have to vote, I have decided, as your independent Chairman, I will accept past practice which was carried out at the last Conference, and I will count the votes as "one man, one vote." In actual fact, all I am saying to you is that I will maintain the status quo as far as voting is concerned.

MR. SMITH (Mr. F. L. Smith, M.C., Hereford Herd Book Society, Great Britain): Did I understand you to say if this matter is re-opened, no further amendments or propositions can be put than the ones already on the book?

CHAIRMAN: I am sorry if I gave you that understanding, Mr. Smith. If this matter is re-opened, you can move as many amendments to the resolution on the book as you like, as long as we get finished by 11:00 p.m.

SR. IRAOLA: Yesterday you told me that I could not vote for Uruguay, but that I had two votes. Why, now, can I have only one?

CHAIRMAN: Because the matter of voting, under the Rules as we considered them yesterday, has been laid aside. It was not cleared up yesterday when we were discussing the Rules. It has been laid aside for consideration by member Societies and recommendations back to the Secretary-General. Therefore, I am forced into the position of reverting back to what was the status quo at the last World Hereford Conference.

SR. IRAOLA: I understand that, but I think the recommendation is for the Sixth World Hereford Conference and not this one. When we began with point number one I had two votes, and when we get to point number two I have only one.

I do not understand your point of view. If we begin with two votes, we must finish with two votes.

CHAIRMAN: All I can say to you is that, in trying to be completely independent and completely fair to everyone concerned, I was under the impression that at the Irish Conference everyone had two votes, but I found yesterday my impression was not correct. Therefore, I must revert to what was the practice at that particular Conference.

Now, is someone prepared to move that the resolution which was passed under Agenda Item 4 be recommitted.

MR. DUEMELAND: A question— it was at our suggestion yesterday that we lay that section aside and it referred to one or two votes; if it would help the Conference to determine this, help our Chairman, it would be perfectly agreeable with me that that be acted on here and now, instead of laying it aside; would that help? I think it is very important that we should treat our members who have come so far with understanding and continuity from what they understood that this Conference meant.

CHAIRMAN: I will ask the meeting, Mr. Duemeland, if they desire to do this. Does the meeting desire to define the voting rights of the countries represented here before we re-open Agenda Item 4?

MR. HUMPHREYS: We would be prepared to take your ruling.

MR. SMITH: We would be prepared to accept your ruling.

SR. E SILVA (Sr. Joaquim de Carvalho e Silva, Associacao Portuguesa De Criadores De Gado Hereford): We will accept the "one man, one vote." Spain has asked me to indicate the same.

MR. HAWKINS: We would be quite prepared to take your ruling.

MR. SWEET: We are prepared to take your ruling.

MR. FIELD: We are prepared to take your ruling.

MR. HAMBLY (Mr. A. E. Hambly, Hereford Breeders' Society of Southern Africa): We are prepared to take your ruling.

MR. COULTES: We are prepared to take your ruling.

CAPTIAN PURDON (Captain D. J. D. Purdon, Irish Hereford

Breeders' Association): We are prepared to take your ruling.

MR. SCAVENIUS (Mr. C. C. Scavenius, Scandinavian Hereford Breeders' Association): I am prepared to take your ruling.

MR. HOUSE: If it is limited to the meeting in hand and has no implications as to the Rules that might be adopted in future, it is very obvious to us that the meeting would prefer to continue as you have ruled, and we would be forced to accept.

CHAIRMAN: The ruling that I have given at the moment will have no bearing whatsoever on the adoption or alteration of the Rules with which you are going back to Member Societies. Will someone please move that the resolution moved by New Zealand, and seconded by Australia, be recommitted?

MR. COULTES: I move, "That the matter be re-opened."

MR. BRADSTOCK (Mr. T. F. Bradstock, M.B.E., Hereford Herd Book Society, Great Britain): I second.

CHAIRMAN: I will ask you to vote.

[It was resolved: That the matter be re-opened. 18 votes for and 3 votes against . . . .]

CHAIRMAN: We now have before us this resolution:

"Be it resolved that the matter of procedures for registration is essentially a domestic matter, but that the World Hereford Council finds no fault in the procedures adopted by the American Polled Hereford Association."

Now we have handed up to us two suggested resolutions.

MR. SWEET: It has become clear to all of us, and as indicated by our Secretary-General, that we are not dealing with membership requirements, nor integrity of records, and that, in his opinion, we are dealing essentially with a domestic matter which is the primary concern of the Societies involved.

On behalf of the American Polled Hereford Association, we would extend our apologies to the Conference, and indicate a high degree of embarrassment for having to bring our domestic linen to be laundered in front of the Conference.

Although tragic it may seem, there is another point of view which we must recognize and appreciate, and that is that this issue is the result of progress. Who would have thought fifty years ago that there would be half a million Herefords recorded each year in the entire world, let alone in one country as the United States? Who would have thought one country would be sponsoring more than three thousand major events in one year? Who would have thought that any country could ever have seventy-five thousand breeders of Herefords—horned and polled?

It is obvious to us, Mr. Chairman that regardless of any action taken at this Conference, the issue cannot be dealt with and the problem solved here today, because it affects the feelings of countless thousands of breeders who are not with us today.

Mr. Chairman, I would offer the following motion;

"Be it resolved that the Conference recognize the full membership status of both the American Polled Hereford Association and the American Hereford Association with full membership privileges, and be it resolved that a meeting be held between the Societies involved, with the Secretary-General of the World Conference in attendance, and a written report be made to the other member countries of the result of that meeting."

"Inasmuch as the Secretary-General will be in the United States during the last week of April, it is suggested that this be considered a possible date."

Gentlemen, this will indicate to you, on behalf of the American Polled Hereford Association, our unanimous desire, our intention to arbitrate any differences that we might have on a domestic basis with our good friends and co-workers the American Hereford Association. Thank you.

CHAIRMAN: I will accept that as an amendment to the resolution now before me as was originally on the books on Tuesday.

SR. E SILVA: I second.

CHAIRMAN: Is there any debate?

MR. MATTHEWS: To enable you to debate this resolution rationally I feel that, as mover of the original resolution on behalf of New Zealand, I should at this stage withdraw the original resolution in favor of the one that has just been put in front of you by Mr. Sweet of the American Polled Hereford Association, and I would ask your permission to do this, Sir, and also ask the permission of the seconder of our resolution on Tuesday.

MR. HAWKINS: As seconder, Mr. Chairman; Mr. Matthews, it is with great pleasure that I withdraw, with you, the motion we proposed on Tuesday. I would like to congratulate you on the resolution that has been put forward, I think it is a good one, and will create much more unity than the one we were proposing.

CHAIRMAN: I must put this to the meeting.

MR. HOUSE: The delegate from Australia has indicated this is a new resolution, not an amendment to the other, is he correct?

CHAIRMAN: At the moment it is an amendment to the other, until such time as this meeting allows the mover and seconder to withdraw.

I must ask the permission of the meeting, Mr. Matthews and Mr. Hawkins, to allow you to withdraw.

Have New Zealand and Australia, as the movers and seconders, respectively, of the original motion, permission to withdraw their resolution? Would all in favor please hold up their right hands?

On a show of hands there were 20 votes in favor.

Permission granted.

CHAIRMAN: New Zealand and Australia are now given permission to withdraw the resolution.

Motion withdrawn.

CHAIRMAN: I now have before me a resolution moved by Mr. Sweet, of the American Polled Hereford Association, and seconded by Sr. e Silva, of Portugal.

MR. HOUSE: I would like to speak to the resolution as it is now on the floor. We have studied this matter and I have heard various statements of the delegates from all the other countries about not interfering with other Societies in their domestic problems. In light of this, I would like to move at this time to table this motion, and I do it because I feel I have in my hand—if permitted to propose—a resolution that we can take home to our folks, and I feel that it expresses the desire of this entire group, if the statements they have made on the floor are sincere. If this the present motion before the Chair is tabled, and my motion is adopted, I feel that we can continue to work within the framework of the World Hereford Council. Therefore, I move that this motion be tabled.

CHAIRMAN: Which motion are you referring to?

MR. HOUSE: The one before the house at this time. This is the only one I could move to table—the one before you. I am simply asking previously to be heard with a resolution, at another time before this meeting dismisses, which I feel will solve some of our problems and I will be prepared to read it at this time before you consider the motion to table, if the Chairman and the meeting desires.

CHAIRMAN: Mr. House, I think you are referring to Mr. Sweet's motion which I have before me at the moment. You want it tabled and not discussed at this stage?

MR. HOUSE: That is right.

CHAIRMAN: Is there a seconder to Mr. House's motion?

SR. BACA: I second.

CHAIRMAN: I will ask you to vote, gentlemen.

MR. HAWKINS: Could we clarify by saying that, by voting, this motion will be tabled and not discussed? Is that what we are voting?

CHAIRMAN: Yes, that the resolution before the Chair will be tabled only. Is that clear to everyone? I will ask you to vote.

[On being put to the vote this resolution was lost with 8 votes for and 15 votes against.]

CHAIRMAN: I am sorry, Mr. House, the resolution to table the motion is lost. The resolution is still before the Chair [Mr. Sweet's resolution].

MR. HOUSE: Could I move a substitute motion, and I would like to read it?

CHAIRMAN: You can move an amendment, which I think you mean by "substitute".

MR. HOUSE: Whatever your ruling might be, I will accept.

CHAIRMAN: Thank you.

MR. HOUSE: In the light of all that has been said, this is the one thing that we have to go home to our folks with, so that it clears the air as to whether or not the World Hereford Congress is trying to interfere with the domestic affairs of the American Hereford Association, and this is the statement, the resolution that we would like to have, and I so move;

"That members of the World Hereford Council resolve: That the actions of this Conference shall in no way be interpreted as an attempt to interfere with the historical and traditional rights of the American Hereford Association to register, service, and promote Hereford cattle produced within the United States, both with and without horns."

Now, I want it understood that does not say "exclusive rights," and it says our "historical and traditional right."

"Nor shall any such action be interpreted as abrogating covenants entered into between the American Hereford Association and the Associations of neighboring countries."

And I simply want to impress upon this Conference that, if we cannot go home with this, a conciliatory statement that you do not intend to interfere with our private business as chartered in the United States, then you simply leave us empty handed, and we feel it is a denial of our rights within the States, or an attempt to abrogate them, so I am quite anxious you seriously consider this one and let us have it at this Conference.

CHAIRMAN: Mr. House, after studying your resolution I feel it is not really an amendment to the one before the Chair, and I will rule that they be dealt with separately, one at a time.

I still have before the Chair the resolution as moved by Mr. Sweet. Is there any further debate? If there is no further debate, Mr. Sweet, you have the right of reply.

MR. SWEET: Mr. Chairman, I do not wish to exercise the right of reply, I think everyone clearly understands the resolution and if there are any further questions and if it is in order to receive any, I would be happy to receive them.

CHAIRMAN: Thank you Mr. Sweet, I shall now put the resolution. You have it in front of you [the resolution having been circulated].

"Be it resolved that the Conference recognize the full membership status of both the American Polled Hereford Association and the American Hereford Association with full membership privileges, and be it resolved that a meeting be held between the Societies involved, with the Secretary-General of the World Council in attendance, and a written report be made to other member countries of the result of that meeting.

"Inasmuch as the Secretary-General will be in the United States during the last week of April, it is suggested that this be considered a possible date."

I shall put the resolution you have in front of you. Will all in favor please hold up your right hand?

[On being put to the vote this resolution was carried with 15 votes for and 9 votes against.]

CHAIRMAN: I have now before the Chair the resolution as read out by Mr. House.

"That members of the World Hereford Conference hereby resolve: That the actions of this Conference shall in no way be interpreted as an attempt to interfere with the historical and traditional rights of the American Hereford Association to register, service, and promote Hereford cattle produced within the United States, both with and without horns; nor shall any such action be interpreted as abrogating covenants entered into between the American Hereford Association and the Associations of neighboring countries."

MR. HOUSE: I so move. I realize some of the things you have done, the gentlemen here feel that they were not interfering, and this is the opportunity to prove to us their intent, and we would hope that this motion would pass without amendment.

CHAIRMAN: Is there a seconder to Mr. House's motion?

SR. BACA: I second.

CHAIRMAN: It is now open for discussion.

MR. SWEET: We are, I guess, the principals in this entire discussion, and I hesitate to rise without indicating our embarrassment again.

In the first place, this was a domestic issue, we considered it a domestic issue, and the only way it ever found its way on to the Agenda for discussion in the first place is that we were requested by our highly respected and capable Secretary-General to place it on the Agenda as a Report, inasmuch as the physical record keeping system had changed. At that time, it was a World Hereford Conference Report, for the information of the World Hereford Council members. It was still a domestic issue at home and was not thrust upon this Conference as a domestic issue.

In our opinion, the World Hereford Council has not interfered in the domestic problems of our country. A simple Report is all that it was classified as, and all it was intended for, and had it gone right through the Agenda and been accepted as a simple Report, we would not be here this morning.

But, Mr. Chairman, in our opinion, the resolution that has been presented is an indictment against the integrity and the honest intentions of this World Hereford Conference, and we cannot in any way interpret the actions of this Conference as an interference in the domestic affairs of our country.

The resolution that has just been passed has indicated that you feel our capable Secretary-General might act effectively in the capacity as an arbitrator, that you could be assured that the discussions take place on a high level, and give his assistance in resolving any problems.

CHAIRMAN: Is there any further debate? [No response.] Mr. House, do you wish to reply?

MR. HOUSE: I want to reiterate that this will clear the record of the intentions of this Conference if it is passed in its original form, and we can go home and ask our people to work within the framework of this Group.

CHAIRMAN: I shall now put the resolution. Will all those in favor please vote.

[On being put to the vote this resolution was lost with 9 votes for and 14 against. It was resolved: "That the World Hereford Council recognize the full membership status of both the American Polled Hereford Association and the American Hereford Association with full membership privileges, and that a meeting be held between the societies involved, with the Secretary-General of the World Council in attendance, and a written report be made to the other member countries of the result of that meeting.

"Inasmuch as the Secretary-General will be in the United States during the last week of April, it is suggested that this be considered a possible date. . . ."]

CHAIRMAN: Is there any other business which any member desires to bring forward? If not, gentlemen, the time has arrived to close this Conference.

I would just like to have it on record that I, as your Chairman, feel that this is the highest honor that has ever been bestowed on me, to have had the privilege of attempting to run this Conference over the last two days. I trust and hope that you all feel that I have been completely independent to you. I also trust and hope that you all go away from this Meeting realizing that the eyes of the world are looking at us, as the greatest beef breed of the world, completely united.

I am sorry if I have, at some times, not quite made myself clear to various members, but even within the English speaking nations we do have some little opposite types of words which have a different meaning, and sometimes I may have misinterpreted what your intentions were.

I would like to say this to you—in my very short acquaintance with Mr. Morrison in the United Kingdom in 1965, he had no idea that he would have to work with me during this particular period, and through circumstances beyond his control he was delayed some forty-four hours on his trip out—so we were not able to get together before this Conference until 9:30 a.m. on Monday morning, and it completely amazed me to see how two men from different parts of the world just seemed to immediately feel we had known each other all our lives, and we worked together as a team. I am extremely thankful for all the work Mr. Morrison has done, and all the help he has been to me over the last two days. Without him, I would not have been able to have gotten this Meeting to its conclusion in a satisfactory manner, as has been done.

I do thank you, Gentlemen, for your co-operation, for your attention, and for your acceptance of any ruling that I have made, whether it be right or wrong; I thank you one and all, and I do thank you for the privilege of having run this Conference for you.

MR. SWEET: Before we leave, and now that the Conference is concluded, we have had a distinguished gentleman with us (as an observer), who has exercised tremendous restraint in the last few days in keeping himself silent, and he has a pleasant few words he would like to give you. Mr. Pat Connolly from California, our Observer.

MR. CONNOLLY: Mr. Chairman, in the long path of years as I have travelled it, I have never seen a finer, nor more fair man in his deliberations—never in haste and always considerate of the entire subject as it came up; and those of you who know me, know how hard it must have been for me, for two and a half days, to keep quiet; but, I think this Conference should give a vote of approbation to the Chairman for his most excellent handling of all these difficult situations, and to your most excellent Executive Secretary and his entire staff. I want, also, to include the host people—the Australian Hereford Society and the Australian Poll Hereford Society, for their hospitality, their graciousness and their lovely treatment of all of us, particularly those who have come a long, long way. Mr. Chairman, I would like to move that it be put on the records, and I would like to have the unanimous consent of this Body to allow it to go as I have stated it.

MR. HOUSE: Having lost every motion that I proposed, I feel no-one is in a better position to commend you and recognize your complete selflessness in your rulings. I would like to second Mr. Connolly's motion and give you to understand that we recognize your difficult position and appreciate totally your attitude. Thank you.

[Carried with acclamation.]

The Conclusion of the Matter

Mr. Morrison traveled to the United States the last week of April, 1968, while enroute home to England. After failing to arrange a joint meeting between the two American associations, he met with the respective Executive Secretaries, W. T. Berry, Jr., of the American Hereford Association and me, representing the American Polled Hereford Association. He strongly urged each of us to encourage within our organizations closer cooperation between the two groups. He emphasized the need for reduced friction between the two and subsequently made such a report to the other members of the World Hereford Council.

The Canadian Hereford Association had aligned itself closely with the AHA throughout the controversy. It was learned that the CHA's major concern was that the AHA would resort to discrimination against Canadian registrations that traced to APHA records. This would mean, simply, that any Polled Herefords bought by Canadians would not be eligible to re-enter the AHA records in later years if resold in the United States. Likewise, none of their progeny in later years would be acceptable to the AHA, whether horned or polled.

The Canadians were reminded of a similar incident several years earlier when the AHA expunged a good number of animals

J. A. "Tony" Morrison, Secretary-General, World Hereford Council, Hereford, England.

that were not acceptable to them causing a severe loss to several Canadian breeders and an embarrassment to the CHA.

It was obvious that the leaders of the Canadian association were anxious to please the AHA leaders. The Hereford societies around the world had revealed their feelings at the conference table and it left the Canadians little choice but to conform to the principles that had been established there.

It was in August of 1968 that the Canadian Hereford Association called a meeting inviting the Secretary-General, Mr. Morrison, and representatives of both the AHA and the APHA to resolve the matter of reciprocal exchange of records. Mr. Morrison and I, as Executive Secretary of the APHA, along with Walter Lewis, President of the APHA, attended the meeting. The AHA did not send a representative. After adequate discussion and explanation to the Canadian Hereford Association Board of Directors, there was unanimous consent to accept the APHA records on a reciprocal basis without reservation. More than seven years has passed since that critical period. Some of the men directly involved in the controversy have passed on. Two of these stayed quietly in the background and played low key roles, but through their personal interest and infinite wisdom had a great influence on the outcome. Tribute is paid here to Sir Kenneth Luke of Australia, General Chairman of the conference, and R. E. "Pat" Connolly of California, an APHA official observer and past president.

Another unheralded hero of the occasion was Walter Lewis, Larned, Kansas, an official delegate for the APHA who is held in such high esteem throughout the world by cattlemen of all breeds.

J. A. "Tony" Morrison, Secretary-General, who termed it as "just doing my job," should be cited for his unyielding determination to steer the council on a steady, straight course and for refusing to allow favoritism.

This complete account of these critical and unusual times has been recorded here so that Polled Hereford breeders of the future can realize that they have a legacy which didn't come cheaply. It was paid for by toil, sacrifice and anxious hours of study and planning on the part of breed leaders.

It has been sixty years since the controversy began and for those sixty years Polled Hereford breeders have endured a restive existence. It took two generations of breeders to grow strong enough to break the shackles of tradition and prejudice. Thus ended the longest cold war in the annals of livestock history. There were times when the future of the breed hung by a delicate thread when indiscreet action could result in a complete loss of breed identity. In retrospect, all that happened was a part of history and there were many contributors to the breed's current happy and profitable existence.

We come to the end of a long and arduous chapter in the history of Polled Herefords, America's first beef breed. We now see the beginning of a new era when the weights that in the past hung heavily on the breed have been laid aside and, now, Herefords without horns can enjoy the same privileges of other breeds and their breeders can concentrate on improvement and promotion.

10. Herd Shot 'Round the World

Early Exportations
Polled Herefords Around the World

Early Exportations

One of the earliest exportations of Polled Herefords from the United States was to Australia. It is interesting to note, also, that the only other organization devoted solely to the recording of Polled Herefords is the Australian Poll Hereford Society.

The first Polled Herefords imported by Australia from the United States in 1920 were two bulls and three females for Henry Beak and Sons and a bull, which later died in quarantine, and a heifer imported by G. H. Horne. The same year Clyde E. Brown, Rushville, Illinois, shipped the yearling bull Gladiator 29th to G. H. Horne, Rockfield, Queensland, and Mr. Gammon purchased a heifer for Mr. Horne. Additional purchases were made by Mr. Horne through Secretary Gammon in 1920 and 1921. Mr. Horne imported more Polled Herefords in 1925. In 1925 Mr. F. Dearden imported a bull, followed by four females in 1924. A total of 25 Polled Herefords were imported from 1920 to 1925.

Resistance in Australia to the new breed was somewhat typical. Breeders were slow to grasp the advantages of a natural polled head and it was several years before the breed gathered full respect and recognition.

After the Australian Poll Hereford Society was formed in 1933, the Victorial and NSW Royal Agricultural Societies were induced to establish Polled Hereford classes in their schedules. Far-sighted Australian breeders meanwhile had started new importations, bringing about 40 head from America and New Zealand. These breeders were Robert Simson, Anthony Horden, James (later Sir James) Sparkes, Louie Leake, James Hanson, R. D. Bryant, J. H. Doyle and the Murdoch family of Wantagadgery. The New Zealand imports came mainly from Wilencote, that country's foundation polled herd.

The first Polled Herefords exhibited in Australia were at Brisbane Royal in 1924 when Mr. A. H. Stirrat exhibited two bulls and two females, and then repeated the performance in 1925. When A. H. Stirrat died in 1926 most of his cattle went to his brother, Mr. J. D. Stirrat, who exhibited in Brisbane that year along with G. F. Horne and F. Dearden. The same three were back at Brisbane with Polled Herefords in 1927 and 1928. Queensland

Class of bulls under 18 months, 1969 Royal Easter Show, Sydney, Australia.

breeders Barton and Elliot exhibited Polled Herefords in 1929 and 1930. The only exhibitor in 1931 and 1932 was Mr. S. A. Plant of Trevanna fame, being joined by Mr. R. E. Dearden (son of F. Dearden) in 1933.

In 1934 the breed made its Melbourne Royal debut with a team of two bulls from L. F. Leake, but the society did not award championships. Polled Herefords were still very much a novelty breed. One of these bulls was Cudgeena Romeo.

Sydney Royal at last opened its schedule to Polled Herefords in 1935 with a showing of three bulls and a female from the Dearden family of Tenterfield. It was here that Cudgeena Romeo took the historic first championship.

At the 1922 association sale at Des Moines, Ia., Secretary Gammon purchased the bull Excellation from Clyde E. Brown for Frederick Dearden, Tenterfield, New South Wales. The following year at the association sale he purchased a number of polleds for the Mt. Larcom Pastoral Co., of Queensland, including Marvel's Dandy, champion bull calf of the show. Marvel's Dandy was bred by R. T. Painter & Son, Stronghurst, Illinois, and sold for $2,000, top price of the sale.

Exports to New Zealand

The earliest record of exportations to New Zealand was in the mid-1920's. However, interest picked up by 1934 when the third exportation of Polled Herefords to New Zealand was made July 28 when William Wescott & Son, Denison, Iowa, shipped from New York the 3-year-old bull, Dale Cutler, Jr., to Miss Bessie Donald, Featherstone, New Zealand. Secretary Gammon of the American Polled Hereford Breeders' Association made the selection of this bull after visiting several herds in the midwestern states.

Dale Cutler, Jr., a double grandson of Polled Cutler G., showed lots of scale and ample depth and width of body and proved a valuable acquisition to New Zealand herds as he was quite unrelated to either bulls or females in earlier shipments to that country.

Polled Herefords Gain Favor in New Zealand

By 1935, the demand became stronger in New Zealand for Polled Hereford cattle, according to the agricultural press of that country, and Hereford breeders found that they were being forced by popular favor to breed cattle without horns. This was made clear by several prominent breeders who bought bulls at F. E. Humphreys' Wilencote sale. Several visitors, as well as a number of local farmers, purchased their first Polled Herefords at that sale.

Some of the prominent breeders made no secret of the fact that they were being forced into breeding Polled Herefords. W. B Whyte, of Hawke's Bay, said that he had been asked frequently if he had Polled Herefords for sale, and when he could not supply them, the Angus breed was turned to in some cases. He bought one of Mr. Humphreys' bulls.

The absence of horns on cattle was a big monetary advantage and butchers in Hawke's Bay frequently gave one pound, or four dollars then, a head more for dehorned beasts than for those retaining their horns. They did this because cattle with horns penned up in the saleyards or in a truck did much damage to hides and flesh.

Eric Beanish of Whana, Hastings, made a similar admission. Beamish Bros., Hereford breeders of long-standing, made a practice of dehorning all their cattle. That was satisfactory from their own point of view, but those who bred from such cattle could eliminate the horns only by removing them, a job that was not relished by some farmers. Therefore, they found it desirable to breed naturally polled beasts in order to supply the growing demand. He also said that dehorned or polled animals were worth more to butchers than horned beasts, and this was frequently proven in the saleyards in Hawke's Bay.

Polled Herefords Around
The World

The longest continuous "trail drive" in the history of beef began on December 17, 1968 (it was only a coincidence that it occurred on the sixty-fifth anniversary of the Wright brothers' first flight). For that drive 275 growthy Polled Hereford calves spritely worked their way up the loading ramp and cautiously stepped onto a $7.5 million DC8 stretched jet on their way to Chile.

The first 28 calves nuzzled each other into compartment number one, just behind the cockpit, to keep the center of gravity forward. The rest began filling the tenth stall in the tail-section compartment, working forward until the plane was loaded.

For this flight many hours and days of research had been done by the APHA staff and special consultants. There were previous cattle air shipments to study. Some had been successful and others failures.

The staff had studied the tragic happening when 30 head being shipped from the United States to South America, improperly secured and tied, shifted forward in the plane. Control was lost, and the plane crashed, killing all aboard.

The staff was also aware of a planeload of hogs en route from Chicago to Europe, that suffocated on the ground in New York. Then there was a load of cattle en route from New York to Europe that expelled so much moisture that it froze on the inner fusilage of the plane, forcing a premature landing.

With these case histories to draw on, it was concluded that the greatest problem in air shipment of livestock was controlling the climate inside the plane during shipment.

All data available convinced the APHA that the "souped-up" DC8 could do the job. However, until the first test flight was completed, there was great apprehension about the success of shipping such great numbers.

Then came that historic day, at Amon Carter Field in Fort Worth. When the hatch was slammed closed and the big jet began to rev up its engines, everyone, to the last cowboy, went to the control tower to hear the pilot's report of the progress of the first few minutes in flight.

The big plane taxied slowly at first to take its position at the very end of the runway to be cleared for takeoff. It made its deliberate turn and headed south, and the great engines began to roar in the distance. It seemed that the giant bird began to groan under its 93,000 pounds of payload and roll forward ever so slowly. The pilot, of course, realized the danger in sudden jerks that could shift calves through their panel partitions and throw the entire load forward,which would spell tragedy. Imagine 275 calves, all 93,000 pounds, slamming against the forward bulkhead and behind the pilot's seat!

The plane slowly picked up speed, faster and faster. One became extremely conscious of how fast the plane was approaching the end of the runway but was still on the ground. A few yards before the plane ran out of airstrip, it seemed that the wheels were retracted from under the plane, leaving it suspended wheel-high in the air. Slowly it inched its way up and finally farther and higher until it became a blur in the early-morning Texas sky.

Those present had been informed that, if there was no drastic change in the temperature or humidity in fifteen minutes, all signs would be "go." After about fifteen minutes the pilot reported that all was smooth thus far. A subdued sigh of relief is the best description of the reaction of those witnessing the world's first massive cattle airlift.

The air shipment of 10,000 head of cattle, all the way to the bottom of the world, 6,050 miles away, captured the imagination of the press around the world. Never before or since, has a beef-cattle activity been so highly publicized or widely circulated. Little known to the cattlemen of the country was the planning and preparation necessary before such a project could be a reality.

Chile is a beautiful country—much like California—nestled on the western slopes of the Andes Mountains and stretching to the Pacific Ocean on the west. It is only one hundred miles across at the widest point and is separated from Argentina by a range of the most rugged mountains in the world. Known more for its copper mines in the arid northern region, it stretches south for two thousand miles and is dotted with crystal mountain lakes in the southern region. There are many waterways in the extreme south. The largest, the Strait of Magellan, separates Chile from two of its most interesting possessions, Punta Arenas (Magallanes) and Tierra del Fuego.

Legend has recorded that the island got its name when Magellan discovered this shortcut to the Pacific in the year 1522. He sailed through the strait late in the evening and saw the fires of the Indians on the beach on the south and called it Tierra Del Fuego (Land of Fire).

APHA representatives first visited southern Chile in March, 1966. Jim Gill, president of the APHA, and I, as executive secretary, made a trip there, sponsored by the Foreign Agricultural Service.

Both of us were greatly impressed by the potential of the area for economical beef production. The area was relatively disease-free and had a mean temperature of about 50° F. Grass was in abundance in spite of the close-cropping for many years by several million sheep. The ranchers were ripe for change because they had endured several years of depressed lamb and wool prices.

There was one great stumbling block—capital to invest in seed stock was not available within the country. Chile was waging an economic battle against inflation and losing. Sheep *estancias* had been held by the same families for many years. There was a growing feeling that, unless Chile could increase production of agricultural products and improve its balance of trade, the country would be overrun by communists. Agrarian reform was initiated to break up the large holdings and get the land into the hands of the *campesinos*, or working class.

The first interest on behalf of the Chilean government was shown in April, 1967, when a request to borrow funds for cattle purchases was made through the Corporacion de Fomento de la Produccion (CORFO) of Chile. The request was a more or less open invitation to the major breed associations of the United States and Canada to help establish a line of credit.

The APHA's first attempt to help was with the Export-Import Bank in Washington. The association was quickly informed that credit was limited only to a good-risk customer after the borrower put up 20 per cent of the money. The seller was required to carry 35 per cent of the loan, and then the bank would loan 45 per cent for three years. Clearly the bank was not the place to look for cattle loans when the cattle business had a slow turnover and must have longer financing terms.

I asked the vice-president of the bank on that occasion, "Assuming that this loan is going to be made, would you like for your bank to make it, or would you rather some other government agency be the lending agency?"

Polled Hereford heifers exported to Magallanes Province, Chile, 1969.

Unloading first cattle airlift to Punta Arenas, Chile.

He said, "What do you have in mind, Mr. Sweet?"

I answered, "I'm not sure, but I believe we will get the money somewhere, and I just wondered if you would like to be a part of something that has so much potential to be good?"

Ironically, about four months later, on the same day that the United States Agency for International Development (AID) informed us that our loan request had been approved, the Export-Import Bank called and approved a seven-million-dollar loan for seven years for the Chileans to purchase Polled Herefords.

Probably the most concerted drive on legislators and the greatest positive response was realized when the Board of Directors of the APHA, past presidents and other influential breeders made their planned approach on legislators to obtain credit for the Chilean cattlemen. Each called his senators and key representatives or sent telegrams, while the ingenious Anthony A. "Tony" Buford made a personal call to President Lyndon B. Johnson. An appointment was made at the White House for President Connolly of the APHA and me. Obviously wheels were beginning to turn. We were received by Jimmy Jones, assistant to the President, who listened attentively to our story. We had a story to tell and a challenge worth accepting. We had discovered a virtual paradise for beef-cattle production, a willing group of cattlemen living in a country that to that day had never defaulted on a loan. They needed help in the form of long term loans to import breeding stock from the United States, which had in the past fifty years developed the greatest source of seed stock in the world. If we wanted to share our progress with underdeveloped nations, why not use the great opportunity and help a most deserving country? AID has spread billions of dollars in loans around the world, but it is doubtful that any have so typically served the intent of that agency or more effectively fulfilled its basic purpose.

Of the original shipments of 13,500 head of cattle sent to the Magallanes Province, it is estimated that there are now, seven years later, 40,000 head of breeding-age Polled Herefords, after slaughter of the steers that were produced for that purpose.

The French Story

One of the unusual exportations took place in 1966, when Ray Rowland, of Belleview, Missouri, retired chairman of the board of Ralston-Purina, exported fifteen head to France. To my knowledge this was the first importation by France from another continent of any breed of livestock in recent years.

The cattle were imported by Jean Duquesne, of Rouen, France. He was required by the French government to enter into an agreement denying him the privilege of selling any of the "exotic" Polled Herefords for breeding purposes. He set his objectives to cull rigidly all of his produce and accumulate a high-quality herd, hoping that someday the attitude of the government would change and he would be allowed to sell to other breeders.

To get the most immediate response from his Polled Herefords, he purchased about seventy-five French Salers cows for crossing. He now sells the calves at slaughter time as American-style steaks through his restaurant in Rouen. His program has been a success by any standard, and demand has outrun his supply. It resulted in one of the most successful beef cattle crossing programs yet observed.

The French Salers was an old but little-known breed developed by French farmers in southwestern France. They were selected for many generations for the high protein content of their milk, which was used to make a specialized cheese typical of the area. The farmers practiced milking three teats for cheese-manufacturing use twice each day and leaving one for the calf.

The Salers became a large, rugged breed over the years of selection, highly fertile, good milkers, and good graziers. The rough, mountainous area that was their native habitat provided an environment that influenced the development of large, stout frames and rugged, hearty constitutions.

Since that first purchase of Polled Herefords, subsequent shipments have been made from Dulick Stock Farm, Morgan, Texas, and Falklands Farm, Schellsburg, Pennsylvania.

The French government has now officially recognized Herefords and Polled Herefords as breeds and authorized an official Herd Book.

Early Importers

The following breeders typify the early importers of Polled Herefords in foreign lands. These innovative men have helped to make the Polled Hereford breed a truly international one.

Henry Beak, Australia.   *L. F. Leake, Australia.*

F. E. Humphreys, New Zealand.

Peter Brockelhurst, Rhodesia.

Sr. Eduardo Ayerza, Argentina.

Sr. Carlos Fournier, Uruguay.

Jean Duquesne, France.

Edward Rushmore, Rhodesia.

Sr. José Ruíz Cifuentes, Spain.

Sr. Olympio Gerra, Brazil.

Sr. Roberto Merino, Ecuador.

Maj. G. E. F. North, England.

Oscar Colburn, England.

Mossom Boyd, Canada.

11. Changing Times

Changing Times

Performance Testing Introduces
The Age of Reasoning

With the introduction of performance testing about 1950 and along with it the concept of measuring traits of economic value, an element of practicality began to enter the minds of seed stock producers and commercial cattlemen alike. Initially, performance testing was crude at best and in retrospect may be seen as a promotional and competitive tool to be used against those who had gotten their cattle smaller and smaller, and shorter and shorter, following the compressed fad in the 1930's and 1940's. The beef industry was vulnerable to the exploits of promoters who weighed and recorded mature weights, but were disinclined to place any emphasis on shape or conformation. It was rather crude the way it all began, and shaped up very quickly into a battle between proponents of conformation only against those who were concerned only about weight and size.

Maybe it's the only way things can happen here in America, but through competition people can be sufficiently motivated to initiate new action. Because of lack of communication and cooperation, it was nearly 10 years before the two factions could be brought together and realize that there was merit to the position that both sides were taking and that neither had the complete answer.

We have very little reliable information on the performance of the foundation animals in each of the major breeds. Of the eight or nine basic economic or money-making traits, probably the only record that we have that we could consider reliable is that of mature size. There is very little we know for sure about any of the breeds from the standpoint of reproductive efficiency or fertility. The records of only a few of the foundation animals have been made and preserved for us. There is almost no information concerning weaning weights that would reflect on mothering ability of dams or yearling weights which would relate to early growth rate. We do have some information about longevity or life span, especially in those animals that serve as foundation animals.

We never cease to be amazed at the genetic resiliance of the bovine beast and especially the flexibility that gives it the ability to adapt to environmental and climatic conditions over several generations. When the Longhorn came onto the scene in about the year 1550, it entered the southern regions of the country and adapted very quickly to the arid, dry, southwest where grazing was sparse. It ranged as far north as it could during the grazing season each year and retreated back to the somewhat more comfortable climate during the winter months. There may be a question about changing the general direction of evolution but we have certainly seen examples where it has been accelerated by the influence of man. Polled Herefords fit into the evolutionary pattern of beef cattle because of the dominance of the polled gene. The rate of change brought on by the influence of selection has been dramatic.

No other specie or breed of livestock in history has been subjected to greater stress or so threatened by controversy and overcome greater resistance and ridicule. The breed seems to not only survive and progress, but actually thrive under pressure.

Changing Times

Performance Versus Popularity

It was nearly two centuries before a Texas ranch was put under barbed wire that the first Mexican cow crossed the Rio Grande. Long of horn and leg, variegated in color, and belligerent in disposition, she was a progenitor of the millions of cattle to fatten upon the grasses of America. She provided the genetic base for modern commercial herds. Since that time evolution has been at work. As isolated gene populations of the world have built up and developed their own characteristic traits, breeds evolved. Even more dramatic has been the revolution happening in our time in the past ten years.

The competitive race between breeds and breeders to identify superior germ plasm has developed to a red-hot pitch. True, there is no breed that can today claim superiority in all economic traits, but the breeder to survive will be the one who does the best job of finding the highest-performing animals in the most valuable traits and let them produce all the offspring they can, while identifying the lower performers and reducing the number of offspring they leave. That is the name of the game—increasing the superior and reducing the inferior.

The use of specific measures of rate of growth, composition of gain, fertility, efficiency, and longevity will enhance the rate of progress we make. The APHA's Guide Lines Program is designed to guide the breeder through the steps necessary to make herd progress and increase breeding value. It is a program to encourage the use of tools most effective in bringing about herd improvement.

The breed's pioneers were some of the beef industry's first innovators in selecting for the hornless trait. Today's modern seed-stock producer will be just as innovative in the use of all available tools to hasten herd and breed progress.

Pedigree by Computer

Having dissociated its herd book from the older (horned) American Hereford Association, the APHA began independent operations by computerizing its ancestral records of 3,000,000 polled cattle. Its immediate objective in data processing was to handle a rising volume of work, since 180,000 Polled Hereford calves were being registered each year and 100,000 transferred in ownership. But, in so doing, the APHA invented something new—a sophisticated method of pinpointing genetic values, called the Performance-Ancestral Record, or PAR Certificate.

Traditionally, a breed association's herd book was simply a record of births, like the "begats" in the Bible. When Bert Gammon registered the first Polled Herefords in 1901, he wrote the names and dates by hand in a ledger book resembling the guest register in an old-fashioned hotel or club. In 1968 most breed associations had not progressed much beyond basic pedigree and a fee of one dollar per calf for the registry certificate and for processing the data. The breeder agreed to keep certain simple statistical records. The associations' computers did all the rest.

Standard of Perfection Program

Today, when a breeder believes that an outstanding sire or dam has appeared in his herd, he may enroll it in a supplementary program of the APHA, the Standard of Perfection Program. Here the progeny records of top animals from different herds are compared, and the best performers become eligible for extra recognition as Superior Sires. In effect this program follows the tradition of county fairs, except that the blue ribbon which represents merit in the show ring is replaced by the Gold Seal award, and the award is based upon scientific, computerized tests.

Since the best test of a superior bull or cow is its progeny, the Superior Sire aspirant follows the Guide Lines Program procedure with a third phase for his selected bull. Up to now the purpose of Guide Lines has been to compare animals *within the herd* as a guide to selection and culling. In the Standard of Perfection phase a bull is compared in performance with the superior bulls of other registered herds.

The "reference sires" for this comparison are ten selected bulls that have been put through a progeny test on 1,000 cows and ranked by computer from 1 to 10. Numbers 1 and 2 are rated Gold Seal sires. If a breeder believes that he has a bull in this elite class, he may give it a progeny test in competition with a reference bull.

Let us say that he selects number 6. Using artificial insemination, his bull and reference bull number 6 are progeny-tested on an equal number of cows under matching management conditions. The resulting calves are raised and followed through to slaughter in order to obtain carcass data. The bull's calves must exceed the performance of those of bull number 6 by the same percentage figure as reference bull number 2 in order to qualify the bull for a Gold Seal award.

A minimum of twenty progeny is considered a fair test of the sire. The resulting information is then relayed to the APHA computers. A Superior Sire will be worth much more than an ordinary bull. He is allowed a much wider use by artificial service within the breed, consequently bringing more improvement to the breed and greater rewards to his owner. The Standard of

Perfection Program is the equivalent of the World Series in baseball or football's Super Bowl in the world of fine cattle breeding.

Gathering show-ring data.

Ultrasonically measuring for fat thickness to gather show-ring data.

Wisconsin State Fair, 1968—Polled Hereford Big Top.

Rail judges, left to right, Mans Hoggett, Mertzon, Texas; Les Brannan, Timken, Kansas; Joe Lewis, Larned Kansas.

The PAR Certificate

The accumulated data from the various phases of the herd-management program are brought into single focus by the PAR Certificate. This Performance-Ancestral Record documents the performance of each purebred calf for one year at least and for a longer period if it is allowed to mature into a bull or brood cow. Its own record is backed up with the progeny and produce records of the sire and dam.

Why is this significant? The repeatability of important performance traits is very high. Researchers have found the records of the sire and dam extremely significant in predicting the performance of their progeny. The PAR Certificate represents a new kind of herd management: genetic engineering.

The Exotic Frenzy

A study of the history of other breeds of beef cattle is a study of parallels. They all have conformed to a growth pattern that is typical of the metamorphosis of plants and animals: birth, growth, maturation, and decline. Some, like the Longhorn, for all practical purposes have become almost extinct, others have declined to a level where they no longer make a serious contribution to the beef industry and have become in effect hobby breeds.

All breeds that have risen to a level of significance have done so because they served a need. They came into being because they provided particular traits that complimented the existing cow herds of America. How fast they rose to prominence and how long they maintained a high level of influence depended primarily on the people who bred and promoted them.

Our grandfathers witnessed the Shorthorn take over from the Longhorn at the turn of the century. It was not difficult because of the natural nick of the two breeds. The Longhorn for generations had adapted to the harsh, rugged environment and had learned to survive the long winters without supplemental feed from man's hand. However, the slow but growing affluence of the young nation whetted the appetites of the people for a more tender, flavorful steak, and there was a decreasing need for work oxen.

Evolutionary changes are evident in the conformation of these Polled Hereford bulls. The great sire Victor Domino, calved in 1930, is shown at top, followed by a 1959 champion bull and, on the following page, a 1973 champion bull.

The Shorthorn cross produced an earlier-maturing animal with faster early growth and higher-quality carcass. These complementary traits, along with the fact that Longhorns had no organization or individual to promote their interest or assets made them easy prey for the Shorthorn onslaught.

Big-moneyed interests of the East and import promoters found a ready market for all bulls brought in from Scotland. Consequently, Shorthorns began to dot the ranges of America, and finally the cow herds of the western range became populated with this multicolored breed.

By 1920, Shorthorns had hit their peak, and the American Shorthorn Association reported a record number of registrations. The next ten years were difficult for all breeds in America, but especially so for Shorthorns. They declined in numbers to scarcely half their number ten years before.

It was during this period also the Hereford breed not only held its own in spite of the rigorous effect of a long depression but became stronger as a breed and began to take over ground lost by the Shorthorn.

Again we might observe a young, opportunist breed exploiting the natural good traits of an already existing, populous breed and complementing them with a few of its own. However, with the advent of the Hereford was ushered in a new era in breed promotion. When the Hereford breeders banded together in 1881 to form their organization, it was for the purpose of promoting and popularizing the breed. Up until that time the beef-cattle world had never seen so dedicated and determined an effort with one single objective. The early Hereford breeders were motivated by an almost fanatical zeal to put a white face on every calf born in America. They came nearer to accomplishing this objective than any other breed.

Hereford breeders had economic assets to promote. Mother Nature teamed with them to bring on a series of hard, harsh winters that proved the hardiness and endurance of the white faces. The breed's ability to withstand the severe winters of the western ranges and adapt their body needs to drought and periods of short feed supplies and still raise thrifty calves was an asset unparalleled by any other breed.

Beginning in the mid-1940's there was the meteoric rise of Angus. A most effective promotional program was responsible for the great acceptance and growth of Angus. The American Angus Association effectively spotlighted the weaknesses of Herefords and played up the strong points of the Angus. The AAA was aggressive in presenting the assets of its breed and successfully launched a program sloganed to be effective. The chant was "Paint the West Black." Ease of calving, elimination of dehorning and mothering ability were attributes of the breed that were much talked about. However, while progress was being made in these traits, performance testing came along, which placed great emphasis on weight gains and lean meat per day of age.

The Angus surge in numbers, while impressive for a period of time, began in the early 1960's to give ground to the exotic frenzy that has typified the last decade. Charolais cattle had been in the United States for many years. Their growth had been rather slow and deliberate. The performance test had, with its emphasis on 140-day feed tests, concentrated much attention on the growth trait, and at this point the beef industry was ripe for exploitation by another breed. The vacuum was developed, and Charolais were there to fill it.

It is interesting to note that everyone was his own expert during this period, which could best be described as the "exotic sixties and the psychotic seventies." Cattlemen were in a mad scramble to get something new to offer, while the research stations were testing every extreme in a desperate move to stay ahead of the cattlemen. Many assumptions were made before data had been collected and analyzed. Magazine editors were irresponsibly reporting every rumor and presumption under the guise of "interpretive writing." Cattlemen would select one breed of bull to improve one trait in his herd and find he had lost ground in one or more others. Many resorted to the use of large, horsy bulls to improve growth rate, only to find they had brought about the more serious problem of calving difficulty in their herds. Others suffered severe loss in quality grades in order to gain in growth rate.

The most serious effect, however, of the exotic frenzy was the enlargement of cow size without regard to efficiency and cow-herd maintenance cost. This did not really begin to dawn on the cattlemen until the harbors around the world were opened to receive the feed grains produced in America in 1973. When the cost of feed grain tripled and the price of fed beef dropped by 30 per cent and feeder calves by 50 cents per pound, the feed-lot operator began to measure feed conversion and cost of gains as he never had before. Never in the history of the cattle business had so much happened in so short a time span. In 1973 feeder calves brought 70 to 80 cents per pound; in 1974 they had dropped to 25 to 30 cents per pound.

The cattleman has become extremely cost-conscious as he watches his cow herd consume high-priced feed. Many factors have contributed to the beef crisis of the mid-seventies, not the least of which was a government administration influenced by short-term political goals, consumer boycotts, price controls, truck strikes, and numerous administrative indiscretions. As if these outside factors were not enough, there was a reluctance on the part of the feeder and beef producer to recognize many signs of pending surpluses. It required only a slight obstruction in the supply pipeline to send the beef market into a tailspin.

In spite of all the problems and negatives of the beef crisis, there is obviously some benefit to be gained. Many herds will get a good culling, and the cattleman will once again look at his cows as a beef factory that must be efficient. The cow must be measured in terms of cost of production and value of output. The beef crisis was the end of an era of exotic frenzy and the beginning of the age of reason and realization in the beef business.

Artificial Insemination as a Tool

The breeder's changing attitude toward the use of artificial insemination has been interesting to follow. Scarcely twenty years ago artificial insemination was considered experimental and expensive.

Many reasons were conjured up to discourage its use. The breeder's subconscious fear, however, was of decreasing calf crops due to poor conception and reducing the need for bulls, which would affect sales. Overreacting on the other extreme was the semen salesman, who promoted AI as a panacea correcting most of the cattleman's problems. While promoting it as a tool to speed up genetic improvement, the selection of bulls placed in the stud for AI use left much to be desired in assuring genetic advancement.

High-powered, well-financed promotional programs were used to encourage the use of bulls that had an opportunity to have great influence on a breed. Those breeds with a small population or narrow genetic base were particularly vulnerable to this promotion. As selection practices of AI studs became more reliable, this danger subsided somewhat. However, AI in beef herds became particularly costly when the conception rate was low and cows missed having calves. There were obvious reasons for cattlemen to approach AI use with a degree of skepticism and caution.

A great deal of progress has been made in the management of AI studs in recent years. Breeders have greater confidence in their ability to provide services and breed improving germ plasm. There is a need for continued guidelines in selecting superior sires that are used widely throughout the industry and especially within breeds. Without established procedures for qualifying a sire for extensive use, a breed may be vulnerable to the exploitation of promoters who are motivated more by short-term profits than by long term genetic progress.

Based on these premises the APHA selected a study committee made up of beef industry reseachers, artificial insemination specialists, and cattlemen. As a result of their study and recommendations the Board of Directors approved the present rules and presented them to the membership. The program allows AI use of young bulls with a record of performance. There is a provision for an increasing amount of AI use, depending upon the level of performance attained by the individual bull. When a bull has attained Superior Sire status, he has the privilege of unlimited AI use. Superior Sires are those that have, through superior performance of their progeny, proved their capability to contribute to genetic improvement in the economic traits and to be free of abnormal genetic lethal traits.

Polled Hereford breeders can view with pride the past seventy-five years since the birth of the breed. They can credit their success to an attitude and a spirit on the part of the breedership that was receptive to change and adapting to change when necessary. The willingness of the average breeder to overlook little things and emphasize the economic traits that affect profit has typified the breed throughout these years.

Although the cosmetics of color pattern are important to maintain a breed trademark, the tendency to argue over a tick of white and create big issues by legislating rules governing breeding programs has been minimal.

Polled Hereford breeders in general have felt that the biggest difference in a great show bull and a good commercial bull should be only a bath. As we analyze and observe the Polled Hereford breed and the people who have produced it, we see a people that have transmitted an image, a people who accept systems that are progressive, and who are willing to change when the need is obviously there.

The future of beef cattle offers great challenge. There will be more interesting innovations in the future—ova transplant, semen sexing, and genetic engineering, in addition to more accurate ways to evaluate carcasses and efficiency, just to mention a few. One of the more revolutionary ideas to purebreeds may be the initiation of sound systems of genetic planning to introduce germ plasm from outside the breed to speed up genetic change.

These are a few of the possible techniques and tools of the future that breeders may adapt for their own use. The breeders who are alert to change, come to grips quickly with new challenges, and grasp firmly the tools of progress will enjoy the rewards of a prosperous future.

Changing Times

Back to the Basics

There has never been a period of time in the history of beef cattle when there was more controversy and differences in opinion than there is today. Researchers and cattlemen are questioning old established concepts. The traditional show ring is being indicted. Beef as a product is under attack. In the last ten years we have experienced a revolution in our thinking about growth rate. The cow has been singled out as the culprit in a world grain shortage. Conventional breeds have been shaken out of their lethargy in the past decade, aroused from their comfortable, smug, feeling of security by the exotics and forced to take a look at time-honored traditions, prejudices and biases.

We have taken a good look at our breed, Polled Herefords, and compared it with competitors. Surprisingly, when we look at it with candor we like most of what we see. How lucky we are to have maintained the most valuable economic traits. When we consider the stress to which Polled Herefords have been subjected, we may conclude they are great *in spite* of what we have done, rather than *because* of what we have done to them.

Most of the traits in which we take great pride, and those which give Polled Herefords the breeding value that we recognize today, were the result of natural selection on the range over a period of many, many years. There is a certain amount of luck involved, also, in the accumulation of the important basic economic traits, such as reproduction efficiency, mothering ability and adaptability. It goes without saying that animals that have low fertility do not reproduce themselves very well. And those that have high fertility can reproduce and leave more highly fertile individuals within the population than can those with low fertility.

I recall the plight of a once great breed that was popular throughout the United States and the world. Its long-time secretary, in answer to my question about the cause of the breed's decline, said, "This breed was great when the cows had a calf every year, gave enough milk to raise it and they grew fast enough to make a profit. They fell from favor when they failed to calve every year, didn't give enough milk to raise their calves thriftily, and they matured too early."

So, whenever we think of some of the things that the Polled Hereford breed has been subjected to that have been counter-productive and have not improved the breeding value or genetic merit, then we realize how lucky we are.

The Way to Extinction

The term "hot-house" refers to adjusting the environment of animals in order to help them survive and propagate as opposed to letting the animals adapt to their environment.

There is one way to avoid the extinction of cattle that are not adaptable to their environment—you can "hot-house" them. If the feed available isn't adequate, then buy supplemental feed. It only costs more money to maintain them. If the cow won't cycle and breed regularly, call your vet and give her expensive reproductive therapy. If she won't calve naturally, set up an obstetrics barn, heat it and put in closed circuit TV so you can watch her from your office. If she won't give enough milk to raise her calf, buy a nurse cow. It will only cost you twice as much to raise it to weaning. If your bull won't get out and hustle, give him an air-conditioned stall and hire another cowboy to patrol the herd and bring each cow to him.

After doing all of these things, in this day of shortages, high prices and slim profit margins, *if your pampered cattle are not extinct, then you, as a cattleman, will be.* The absurdity of these artificial practices is obvious. The industry can't afford such practices any longer.

There is only one way to endure in the cattle business—it is that way today and has always been that way—"get back to the basics." We are fortunate indeed that only a small percentage of Polled Herefords have been subjected to this kind of care and management while the vast majority has been left to roam free on the range and to develop over several generations and adapt to the climate and the feeding conditions.

Shows Can Have Value

For many years our distorted show ring standards were probably the greatest impediment to genetic progress. You will notice that I didn't say the *show ring* had been an impediment to genetic progress but rather our show ring *standards.* Many of these standards still prevail in show rings throughout the world where they are in violation of every common sense practice and completely out of tune with nature.

On the other hand, we may quarrel with those who fail to see *any* value in the show ring for a breed. Although the value, insofar as genetic improvement is concerned, is questionable, we have failed thus far to find another tool that gives us the public exposure, education and promotion that we find in the show ring. We haven't learned yet how to use the show ring to best advantage, and to gain the ultimate objective of demonstrating improved breeding, squeezing out the utmost in promotional value from public exposure and sharpening it as an educational tool.

It is completely illogical to expect seed stock to thrive efficient-ly on roughage and pasture, to expect steers to have high grading, high cutting carcasses with only a thin rind of outside fat and, at the same time, bring our show cattle into the ring with an inch and a half to two inches of grease on them.

The show ring, if used in a progressive manner, can contribute to breed improvement. We began measuring weight and fat thickness at the National Show in 1969. Since that time we have seen the fat thickness of bulls reduced from .62 inch in 1969 to .23 inch in 1974, while the weight per day of age increased from 2.28 to 2.39 pounds, respectively (approximately fifty pounds per head). This progress has been paralleled on the farms and ranches by higher conception and calving percentage in show cattle and with more of our breed topping competitive performance tests and much greater demand for our commercial stock bulls. So, the show ring can be useful if we use it properly.

The Truth About Grain-Guzzlers

What do we see as the future needs of a grain-hungry world? What is a grain-hungry world going to require of the breeds that survive? The answers are that the world wants to reserve all the human edible grains for human consumption and it wants cattlemen to use our feed grains in the most efficient manner.

There is a shortage of energy in this world and it has hit the livestock industry as surely as it has the airlines, factories and railroads. Energy is the basic stuff that makes the world go 'round. It is the stuff that makes it possible for us to condition our environment and tolerate the extreme heat or cold, whether we are at home, in the tropics or the frigid north; whether we are on a space ship or working in a factory. It's the basic stuff that propels us to work or play and moves the machinery that helps our society function.

Regardless of the many reasons given for the shortage, the

problem is real. It exists and experts concede it isn't going away suddenly or quietly. Energy can be in the form of coal and oil, which is the most talked about because the shortage of oil came on us in a seemingly sudden way—like overnight—and an Arab embargo made it even more dramatic. The Arabs found our weak spot and learned the world would plead, beg and pay excessive prices to avoid walking, riding bicycles or staying home.

We found, however, that we became more willing to get back to the basic economics in the use of energy to make sure we had enough to perform the essential functions. Around the world, man began to look upon cars as essential transportation rather than a luxury. The demand for economy cars that served the basic need for transportation jumped 200 percent. The market for gas guzzlers with expensive accessories dropped overnight by 70 percent.

There is a parallel in the supply of energy for man and animals. It is in the form of food stuffs, grain primarily. An affluent world with rising incomes, greater buying power, along with a change in our national marketing policy, has placed heavy demands on what we erroneously thought was an abundant supply of grain. This poses several questions to the beef producer. Can livestock compete with humans for the available grain? Will the grain-beef price ratio allow the feeding of grain to beef cattle? If so, how much? Will the feeding and management changes which are evident bring about a change in the type of cattle the seed stock producer will need to breed? This time the beef industry cannot be pushed around by high-pressure salesmen who fill the atmosphere with promotional promises and confuse the issues with great claims and few facts.

These are questions that need to be pondered carefully, and answers must result from factual analysis and be accepted with candor. An industry is at stake. Future plans and genetic change, if any, must be deliberate and guided by logic and thought.

During the energy crisis the industrialist has been forced to get back to the basics to measure efficiency in using his available energy for production. The environmentalist has been forced to be a realist. He has learned we can't cap the smokestacks and live in a sterilized world. A certain amount of pollution is a part of nature's process. But nature has a way of taking care of her pollutants.

The cattleman must likewise *get back to the basics* of nature. He can't afford the luxury of a lot of accessories like cows that do not perform efficiently, fail to reproduce and have a high maintenance cost. The basic facts of the cattle business are that the cattleman must have:

1. A calf every year from every cow.
2. A cow that gives enough milk to raise her calf thriftily.
3. A cow with the inherent growth factor that will allow the calf to grow rapidly to market weight and grade at an early age.

These functions must be performed efficiently and can be instilled in the genetic make-up of beef herds by selection. If we carefully weigh the facts that experience and research has given us today, we can do a reasonable good job of *getting back to the basics.*

The basic facts are that big, horsy cows cost more to maintain; 80 percent of the cost of producing a carcass for market is invested by the time the calf is weaned. Calves that grow fast at an early age have higher quality, more desirable carcasses and will bring a premium at the market place.

Cow herds of the future won't be too much different from those of the past. They must be of optimum size and adapted to the environment and feed conditions of the area in which they are expected to function. If they are not adapted in size, hardiness and functional ability, then give them about two generations—

they will either adapt or become extinct. That's a basic law of nature.

Public Image Essential to Survival

No one has a greater appreciation than I for the tremendous contribution that the researchers have made to our storehouse of knowledge the last few years. They have pointed out some things to us, not too diplomatically at times, but they have focused our attention on some real weaknesses in our breed improvement and management programs. That's their job. But at the same time I would like to solicit the same kind of respect and appreciation from the researcher for the cattleman, the breed society and their point of view. I would like for the researchers to consider the responsibilities these cattlemen have, not only insofar as genetic change is concerned, but also their responsibilities for promotion, public relations and breeder services. All of these are absolute necessities for a breed that is going to survive in this competitive world. Our concern is not only for the survival of breeds, but for the very existence of the beef cattle industry.

Our product has been under attack from misinformed and misled medical researchers for many years and is now facing its greatest onslaught from the environmentalist and the around-the-world "do-gooder" who is laboring under the misconception that if he can eliminate beef cattle he will provide the world with more grain for human consumption.

The beef cattle industry around the world faces its greatest challenge from the psuedo-expert in human nutrition and half-educated benevolent groups. How fortunate we are that beef has two great qualities. It is not only nutritious—it tastes good!

Three-Legged Milk Stool

We may liken the responsibilities of a breed association or society that would like to serve its members in the most effective way, to a three-legged milk stool that has a rather delicate sense of balance. One leg might represent breeding value or genetic progress. *A breed society should take advantage of all the factual findings of research and practical experience in helping to develop programs or vehicles that would assist its breeders in making progress in "breeding value."* The breed that does not serve a basic economic need and relies primarily upon promotion, may be like a balloon that is inflated with hot air. It will rise slowly to where everyone can observe it briefly but then, when it reaches an altitude and cooler air, its descent will be faster than its rise.

A lot of drum-beating, horn-tooting, and hoopla without serving a basic need may last for a short period of time, but that breed will finally retreat back to the hands of a hobbyist who can afford to subsidize it.

Promotional Leg

On the other hand a breed might have the most to contribute insofar as breeding value and economic traits are concerned but *unless it is promoted and advertised it is unlikely that it will survive and serve the basic needs of a grain-hungry world.*

"He who has a thing to sell, and goes and whispers in a well, is not so apt to get the dollars, as he who climbs a tree and hollers." It pays to advertise.

I am told there is very little difference in the nutritional value or flavor of chicken eggs and duck eggs, but there is a big difference in the habits of these two fowls. If I asked a group of people how many had duck eggs for breakfast I doubt if anyone would respond. But if I asked how many had chicken eggs for breakfast, we might have a goodly number indicate that they did.

Now what is the basic reason chicken eggs are so much more popular than duck eggs? Let's look at the habits of these two creatures. When a duck gets ready to lay its egg, it takes a good look around to see if anybody is watching. It will find a hedge row, go in one end and, in some secluded spot, scratch out its nest and sit down to lay its egg. After it has completed its chore, it will then go out the other end of the hedge row, to conceal the spot where the egg was laid.

Now observe the hen as she makes quite a fuss about her nest. She selects a place clearly visible to everybody and, without hesitation or embarrassment, settles down in plain daylight to lay her egg. When she completes the job, she jumps up on top of the nest, cackles a few times and fluffs her wings and tells everybody that here is a nice fresh egg for their breakfast. This is promotion in its most natural form.

Without a good sturdy promotional leg under the breed's milk stool, it will lose the balance that is necessary for breed survival and progress.

Breeder Services

In addition to the first two functions of a breed association—to promote genetic improvement and to promote public image—there is the last and very important leg: breeder services. The basic purpose of the organization is to enable people to do things collectively that they could not accomplish individually. It must provide motivation, leadership and tools for merchandising to breeders.

The breed society can provide the exchange of ideas necessary to promote change and progress. It can draw on the experiences of successful breeders and researchers and pass them on to others. It can and should develop a national and international image of a breed that will make it receptive to new breeders and investors, and most of all, present to the commercial stockman an image of "breeding value and profit." And then we arrive at the most important part of a successful breed society. *It must be made up of seed stock producers who are open-minded, receptive to change when it is due and willing to become missionary-minded in their own areas.*

I recall a survey made several years ago to determine what motivated new breeders to choose Polled Herefords. Ninety percent of the respondents said they were influenced most by another breeder. The Polled Hereford milk stool will continue to stand firm if continued emphasis is on improved performance, promotion of the breed image as one that serves a basic economic need, and the association's ability to provide breeders with modern tools for merchandising.

12. Polled Hereford World

The Voice for Polled Herefords

Polled Hereford World

The Voice for Polled Herefords

The first serious attempt at publishing exclusively for the Polled Hereford industry was made by Hiram Herbert, publisher of the *Plantation Stockman* at Montgomery, Alabama, in the early 1940's. He later put a Polled Hereford logo on the front of it and named it the *Voice of Polled Herefords*. Later, in 1945, he changed the name to the *Polled Hereford Magazine*. Hi Herbert, as he was known, exercised unusual editorial freedom and opinion. He traveled extensively, visiting a great number of Polled Hereford breeders, personalized his column and did a great deal of "interpretive writing." At a later date he sold the magazine to Bernie Hart and Associates.

During this time Frank Farley, Sr., a long-time field man for the *American Hereford Journal*, was publishing the *Southern Stockman* in Memphis, Tennessee. A group of breeders encouraged the publishing of another Polled Hereford magazine to compete with the one that had been started by Herbert. Frank Farley, Jr., who at the time was a recent graduate from Kansas State University, along with his father and Jewett Fulkerson, an auctioneer and Polled Hereford breeder, began publishing the *Polled Hereford World*.

Frank W. Farley, Sr., the late editor-publisher of Polled Hereford World.

When Frank , Jr., was drafted into the army, his father sold the *Southern Stockman* and during Frank, Jr's. absence published the *Polled Hereford World.* The owners of the *Polled Hereford World* ultimately purchased the *Polled Hereford Magazine* and combined the two, which resulted in what is known today as *Polled Hereford World*, the official publication for Polled Herefords.

On August 15, 1947, Volume 1, Number 1, of the *Polled Hereford World* was completed and mailed to all breeders of Polled Herefords. Thus began one of the livestock industry's most successful and effective publications. Since that time about 350 issues have been printed and mailed.

No doubt, the founders, Farley, his son, and Fulkerson produced that first issue (like so many other beginnings in the Polled Hereford industry) with some apprehension. There was the formidable *American Hereford Journal*, the former employer of Fulkerson and Farley, which claimed a dominant position in the Hereford world. It was the self-acclaimed official publication of all whiteface cattle, polled and horned.

In the first editorial the editor spelled out the goal of the new journal:

"In the journalistic field, the *Polled Hereford World* is dedicated to Polled Herefords which, along with their breeders, are highly deserving of specialized publication support. The Publishers and Editors of the *Polled Hereford World* welcome the privilege and the opportunity of shouldering this responsibility which they will fulfill faithfully and with fairness to the fullest extent of their ability.

"There is a vast territory for the expansion of all improved beef cattle, and the Polled Hereford will expand in direct proportion to its admirable merits, the aggressiveness of its breeders and the extent to which its complete story is presented to the cattle-raising public of all areas. The latter job is that which the *Polled Hereford World* will do.

Goal and Aim

"You have in your hands the first issue of the *Polled Hereford World*, which we think is a good issue. We truly hope that you think the same thing about it. It contains a vast number of historical Polled Hereford facts that have never before been published in one volume, and which should, therefore, be most valuable as a reference.

"To give faithfully the most cogent publication—service that is possible—this to be rendered co-operatively with all other forces engaged in and for the overall advancement of the Polled Hereford industry—will be the constant mission of the *Polled Hereford World*."

APHA Purchases Polled Hereford World

Eighteen years later on October 20, 1965 I, as executive secretary, received a call from McWhirter Printing Company, informing me that the November issue of the *Polled Hereford World* would not be printed or mailed unless a $7,000 deposit was made immediately. The printer stated that the publisher was $60,000 in arrears on his printing bill and that he was not printing another issue except on a cash basis. He held a mortgage on all the magazine's assets, and, unless something was done immediately, the magazine would become extinct.

The November issue contained the advertising of 163 breeders and the upcoming National Polled Hereford Show story and promotion. It obviously was of great economic value to the breed and specifically to Polled Hereford breeders, whose entire year's production and income depended upon the publication of this issue. Assurance was also needed that it would be a reliable communication organ for the breed in the future.

A conference phone call was made to the Executive Committee of the APHA, and permission was granted to advance the publisher $7,000 to assure the publishing of the November issue, while an emergency board meeting was called to discuss and resolve the problem of future issues.

It was obvious that the publishers had reached the end of their financial rope, and had exhausted their sources of credit. The magazine, for all practical intents and purposes, had fallen into the receivership of the printer, and he was seeking any solution to minimize his losses and be rid of the responsibility of printing it. Although the journal had real value to anyone who would capitalize it with funds to operate and satisfy the creditors, it had actually more value to Polled Hereford breeders. They had much at stake in seeing that it was a well-managed, effective organ to provide a medium for advertising of the breed.

In the emergency meeting of the board two alternatives were considered: (1) to seek a publisher willing to invest in the magazine that had been operating at a loss under the then-current management and (2) to purchase the magazine and manage it as a subsidiary of the APHA.

It was decided to make an offer to the owners that would be fair and equitable to them and also satisfy the creditors. The agreement was concluded on October 27, 1965. The association assumed the liabilities of the magazine, which totaled approximately $67,000, and agreed to pay the owners $60,000—$30,000 in cash and $30,000 over a period of three years. It agreed to retire the outstanding debts to the printer at payments of $1,000 a month. The magazine was incorporated as a taxable, wholly owned subsidiary of the American Polled Hereford Association, under the name American Polled Hereford Publications, Inc.

The first general manager of the *Polled Hereford World* under the new ownership was Louis F. Freeman, and the managing editor was Marilyn Sponsler. Louie Freeman came to the magazine from the advertising staff of the *Farm Journal.* Marilyn Sponsler, with a degree in animal science and agricultural journalism from Iowa State University, continues today as the managing editor of the *Polled Hereford World*.

The growth of the magazine has been phenomenal. In 1965, when it was first purchased by the APHA, the monthly circulation was 7,500. In 1975 average circulation is in excess of 22,000. In the fiscal year 1965 a total of 1,250 pages were printed in the twelve issues, with a gross income of $252,157, while in 1975 2,216 pages were printed, with gross revenue of $687,467.

In 1970 a major change was made in the sales staff of the *Polled Hereford World*. The traditional staff of five field men was combined with the ten existing field men of the APHA, and the areas were redistricted into thirteen areas. Their titles were changed from field men to area activity coordinators. They assumed dual roles and smaller territories. It is estimated that about 30 per cent of their time is spent on magazine duties and about 70 per cent on regular association responsibilities, although admittedly it is difficult at times to draw lines of differences between these responsibilities.

The result of this innovation, which has since been followed by other breed organizations that have their own house organs, is to reduce overlapping visits, reduce duplication of travel, and better compensate each area coordinator for a more effective job for the breed. A major advantage also to the dual role of area coordinator concept was to reduce the size of territories that the men had to cover to shorten travel routes and permit more time at home and fewer nights on the road.

The *Polled Hereford World* has become a pacesetter in the livestock publication field. It has initiated several innovations that have proved effective and economical.

The subscription list is made up of all active Polled Hereford breeders, and many commercial cattlemen, universities, experiment stations, bankers, and federal agencies. Special issues are mailed to all vocational agriculture teachers and county extension directors.

13. Leading the Way

Leading the Way

Association Leaders

The American Polled Hereford Association is a good example of what might be termed "democracy at work." The leadership consists of a board of directors of twelve men elected by the active member-breeders. The twelve districts in the United States elect by popular ballot their own national director, who serves for four years.

The American Polled Hereford Association has traditionally sought ways to utilize to the best advantage the unique talents of its national directors. The members of the board are typified by a sincere desire to serve the best interests of the breeders they represent and almost always place their own self-interests second to that of the association they proudly represent. All national directors serve without pay.

The directors elect from among themselves a chairman and a vice-chairman each year. The vice-chairman is usually a director serving the second year of his term and serves as chairman during his third year.

Early in the history of the association the elected leader of the breed for that year was called the president. In 1971 a new corporate structure was adopted by the board and ratified by the membership providing that the chairman of the board was the chief elected officer. The title president was then transferred to the chief administrative officer, who had previously been called the executive secretary.

For many years the chairman of the board traveled extensively at his own expense, representing the APHA at various cattle functions and area and state Polled Hereford gatherings.

It is interesting to observe over the years how well the individuals elected to fill this important position of leadership were uniquely qualified to solve the problems during the times that they served. A great deal of respect and credit is due the men who were elected to serve the association in this important capacity.

The following pages are devoted to the breed's leaders who served as president, or what is now termed chairman of the board, from 1909 to the present time.

J. E. Green, Muncie, Ind.
1909—16

E. H. Gifford, Lewiston, Neb.
1916—17

A. L. Duncan, Seaton, Ill.
1917—18

P. M. Schooley, West Liberty, Ia.
1918—19

N. M. Leonard, Waukee, Ia.
1919—20

John Herold, Lewiston, Neb.
1920—21

H. N. Vaughn, Stronghurst, Ill.
1921—22

W. A. Wilkey, Sullivan, Ind.
1922—23

George T. Rew, Silver City, Ia.
1923—24

C. E. Brown, Rushville, Ill.
1924—25

P. S. Kendrick, Albany, Tex.
1925—26

H. L. Schooley, West Liberty, Ia.
1926—27

Boyd Radford, Newark, Neb.
1927—28

F. O. Peterson, Galva, Ia.
1928—29

H. J. Smith, Bellwood, Neb.
1929–30

P. M. Christenson, Lone Rock, Ia.
1930–31

Fred W. Schnoor, Perry, Ia.
1931–32

W. H. Campbell, Grand River, Ia.
1932–33

Lester Curran, Mason City, Ia.
1933—34

R. C. Glaves, Lewistown, Mo.
1934—35

Arch R. Dunbar, Des Moines, Ia.
1935—36

Lewis Johnson, Jacksboro, Tex.
1936—37

J. B. Shields, Lost Springs, Kans.
1937 (March-August)

Everett Hodgson, Ottawa, Ill.
1937—38

Frank L. Robinson, Kearney, Neb.
1938—39, 1947—48

Henry Worner, San Jose, Ill.
1939—40

Clifton Rodes, Louisville, Ky.
1940—41, 1942—43

M. P. Moore, Senatobia, Miss.
1941—42, 1948—49, 1957—58

J. E. Lambert, Darlington, Ala.
1943—44, 1955—56

Jim Gill, Coleman, Tex.
1944—45, 1965—66

A. G. Rolfe, Poolesville, Md.
1945–46

Mans Hoggett, Mertzon, Tex.
1946–47

John M. Lewis, Larned, Kans.
1949–50

Adna R. Johnson, Clarkesville, Ohio
1950–51

A. B. Freeman, Walls, Miss.
1951—52

John Trenfield, Follett, Tex.
1952—53

Robert A. Halbert, Sonora, Tex.
1953—54

John H. Royer, Jr. Glenwood, Md.
1954—55

John Shiflet, Red Rock, Okla.
1956—57

D. C. Andrews, Union, Mo.
1958—59

P. H. Ginsbach, Dell Rapids, S. D.
1959—60

W. P. Morris, Jackson, N. C.
1960—61

K. P. Gatchell, Columbus, Miss.
1961—62

Howard Marks, Tracy, Calif.
1962—63

R. L. Swearingen, Sr., Reynolds, Ga.
1963—64

George Kemnitz, Perry, Okla.
1964—65

R. E. Connolly, St. Helena, Calif.
1966—67

Walter M. Lewis, Larned, Kans.
1967—68

L. W. Storm, Dripping Springs, Tex.
1968—69

Leon Falk, Jr. Schellsburg, Pa.
1969—70

Chester Gullikson, Bath, S. D.
1970—71

Harold Hunter, Stillwater, Okla.
1971—72

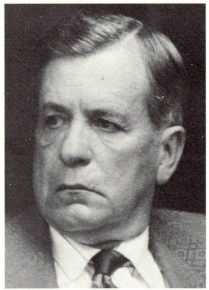

Harry V. Dulick, Morgan, Tex.
1972—73

G. C. Palmer II, Charlottesville, Va.
1973—74

Leland Herman, Wayne, Neb.
1974—75

Lloyd Clarkson, Winfield, Kans.
1975—76

Leading the Way

The Hall of Fame

The Polled Hereford Hall of Fame is unique among purebred livestock registry associations and has become one of the most impressive sights for visitors to the national headquarters. The glittering, impressive display of bronzed hats actually worn by the breeders are enshrined there, and the plaque pictures memorializing them have created an unusually historical atmosphere.

The American Polled Hereford Association Hall of Fame had its beginning in 1965, when B. O. Gammon was presented the first bronzed hat and plaque picture. In the following year six more of the breed's patriarchs were added. These pioneers were Warren Gammon, William Spidel, Robert Halbert, Earl Blanchard, Frank L. Robinson, and John M. Lewis.

There were other giants in those days (besides Giant the fountain head sire of the breed). The pioneer breeders were giants when it came to meeting adversity with resolve, discouragement with optimism, depression with determination.

A review of the attitudes of these men convinces us that, if the heart is right, one can succeed in spite of obstacles. They were rugged individuals in those tough days, when the very existence of the struggling young breed hung in the balance. Every year seemed to bring on a new crisis; and if these natural problems were not enough, there was the ridicule and derision from the horned breeders.

It was along about National time each year that these pioneers would gather to reassure each other of the path they had chosen to tread. In those days breeding and selling Polled Herefords perhaps required a bit more of the individual than today. It was a gutty group of renegades who dared to challenge the "proud and the powerful" of the beef breeds in competitive battle. Oddly enough, most of the rocks were thrown by "close kin" that would have greatly benefited by the success of the polled segment.

In the depths of a cattle depression in 1922, B. O. Gammon's report at the National Sale went like this:

"Attendance records were not broken, price records were left unscathed, sale averages were much below the peak of former years, yet the seventh annual Polled Hereford Week at Des Moines, Iowa, Jan. 30-Feb. 1 must be set down as really the biggest success yet attained by the supporters of the muley Herefords. Why? Because there was evidence from the first hour of the occasion until the last visitor departed, a spirit of 'breed confidence,' which in view of what breeders have endured in recent months was nothing short of marvelous. Not a single 'croaker,' not a man who talked 'quit,' not a single note of pessimism, not a man but was facing forward and confidently expecting better days."

Those were the days without promotional programs, performance programs, pension plans and coordinated activities. It required weeks to travel to a National show.

But the faithful would plan the long trek to Des Moines, spawning ground and the birthplace of the polled idea. Each year with their emotions recharged and bolstered with renewed confidence, they would return to the ranches and farms. They left sharing common goals, dreams and aspirations of some day *not* becoming the world's number one breed but just being a respected partner in the world of beef cattle.

Well, old-timers, wherever you are, it's been a long time coming about, but we all hope you are aware of the results of your efforts. We see the fruition today from the seed you sowed so many years ago. *Polled Herefords today stand as the most promising breed in America and are having the greatest impact on modern beef production worldwide of any breed.*

Although we have problems today, we have better tools to solve those problems. Today's generation silently asks itself, "Under the same conditions, could we stand in their shoes and wear their hats?"

It is for such as these that the Hall of Fame was conceived to perpetuate the spirit that was responsible for their keeping the breed alive during adversity that the world may benefit today.

Remember this in the days ahead—motivation and morale are everything.

The following pages are devoted to Hall of Fame members, led by Warren and B. O. Gammon.

Warren Gammon   *B. O. Gammon*

John M. Lewis

Robert Halbert

Frank L. Robinson

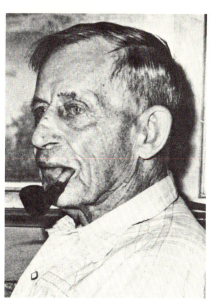

William Spidel

John M. Lewis

John M. Lewis was eighty-four years of age when he was inducted into the Hall of Fame. He was a past president of the American Polled Hereford Association. He was born in 1882 in Illinois. He was a veteran at sixteen years of age of the Spanish-American war (he said that he joined the army to get out of school). He moved to Kansas in 1909. In February, 1910, he married Bessie Libby. Her father was a charter member of the American Hereford Association. He offered his daughter one thousand dollars or ten horned heifers and a Polled Hereford bull as a wedding gift. She chose to take the ten heifers and Polled Hereford bull. The heifers were mostly Anxiety 4 breeding. She and her husband later bought ten more heifers from Libby, and that was the basis of their Polled Hereford herd.

Lewis bred many famous bulls that have left an indelible mark on the Polled Hereford breed. Some of them were Victor Domino 126, ALF Beau Rollo 11, and ALF Choice Domino 6th. Lewis was also responsible for founding several prominent cow families that have become breed landmarks.

Lewis' operation grew from 160 acres to 3,400 acres. He was nicknamed "Peanut John" by his sister-in-law because he always bought his wife a sack of peanuts to eat while he talked Hereford cattle with her father rather than take her out. The first big purchase he made from his money from his cattle sales was a Model-T Ford. He was a great storyteller and had a tremendous memory for details of the past. He and his sons, Walter and Joe, have produced many national champions, and one of his greatest thrills was to exhibit the 1946 National Champion bull, which sold for a record $35,000.

Lewis was inducted into the Hall of Fame in 1966.

Robert Halbert

Robert Halbert was born in 1891 in Sonora, Texas. He started in the cattle business in 1917 with horned Herefords, and with his initial group of cattle he also purchased three Polled Herefords, with the result that he entered the Polled Hereford business. He credits much of his success to his very early start in the Polled Hereford business from using a bull named Domestic Mischief and selecting for thickness and doing ability.

Polled Herefords have been the lifelong work—and conversation—of Robert and his wife, Battie. He produced and showed many national champions and served as president of the American Polled Hereford Association. He founded the Domestic Mischief and Domestic Woodrow families of Polled Herefords.

He produced a good number of great herd bulls that have had an influence on the breed. He bred Domestic Mischief 6 and Domestic Woodrow 23. He was a partner and coworker with Mans Hoggett, of Mertzon, Texas, also a member of the Hall of Fame. He later formed a partnership with Lee Fawcett, a son-in-law. In the early 1950's a severe drought forced him to leave his home ranch at Sonora, Texas, and he relocated his herd in southwest Missouri, but at a later date he returned to Sonora, where he still resides.

Halbert was inducted into the Hall of Fame in 1966.

Frank L. Robinson

Frank L. Robinson was eighty-eight years old when he was inducted into the Hall of Fame. He was born in Wisconsin in 1878 and moved to Nebraska in 1909. He started in the cattle business as a feeder and commercial cattle producer and then started in the horned Hereford business in 1922. In 1925 he acquired his first Polled Hereford bull, known as Polled Pride. The bull was purchased from Radford and Son and was crossed on horned Anxiety 4 cows.

He concentrated his breeding program on Lamplighter bloodlines, produced many outstanding herd sires, and was instrumental in establishing many new herds by providing foundation seed stock. He managed his herd with his son and his partner, Les Robinson, until it was dispersed in 1968. Robinson was a president of the American Polled Hereford Association and served two terms on the Board of Directors. He was known for his congenial manner and ardent love for Polled Herefords and was a strong supporter and promoter of all Polled Hereford activities. His contributions have been great, and future generations will reap immeasurable benefits from his dedication and foresight.

Robinson was inducted into the Hall of Fame in 1966.

William Spidel

William Spidel was eighty-five years of age when he was inducted into the Hall of Fame. He was born in 1881 in Kansas and moved to Montana in 1900. His lifelong activities were many and varied. He was engaged in banking, machinery, and Polled Hereford businesses. He got his first start in Polled Herefords in 1923. His first work was as a cowboy with the Courtland Cattle Company of Roundup, Montana. He worked his way up to foreman and then became manager and later bought the operation. That is where he had lived until his death.

Spidel and his wife, Hazel, began with horned Hereford cattle. His cow herd numbered about five hundred brood cows. He also bred horses. He dispersed his herd in 1963 and entered the commercial cattle business with his son, Ed Allen Spidel.

To many of his friends he was the second Will Rogers, always with a ready wit and an appropriate answer. One of his favorite quotations was, "If you breed cattle good enough, a man doesn't have to work. The cattle will take care of him." He phrased a prayer, often quoted by cattlemen, which goes, "Lord, please make my cattle big enough that I will never have to lie about them." His lifelong efforts will be recognized for years to come.

Spidel was inducted into the Hall of Fame in 1966.

Earl Blanchard

John E. Rice

Mans Hoggett

M. P. "Hot" Moore

Earl Blanchard

Earl Blanchard was one of the early pioneer breeders of Polled Herefords. He was seventy-three years of age when he was inducted into the Hall of Fame. He was born in Saline County, Nebraska, in 1893. He moved to Oshkosh, Nebraska, where he made his home, in 1923.

He and his wife, Isabel, started in the cattle business with horned Herefords and registered his first Polled Hereford calf in 1928. It was sired by one of the old foundation Polled Hereford bulls known as Bullion 5. Blanchard never showed his cattle in the show ring. He maintained about four hundred head in his cow herd. Carl Lindgren, his son-in-law, joined him in 1940, and the farm has operated as Blanchard and Lindgren since that time.

Blanchard practiced a line-breeding program and did not add any new blood after 1959. He sold cattle in every state in the United States, including Hawaii, and seven provinces in Canada.

Blanchard's favorite expression was, "This is the way we like to do it." He felt he had attained success because he had gained lifelong ambition of "living by the river and raising cows."

His cattle found their way into the foundation cows of almost every herd in America.

Blanchard was inducted into the Hall of Fame in 1966.

John E. Rice

John E. Rice was born January 15, 1897, on an Indiana farm, the tenth of a family of eight boys and three girls. In 1915 he went west, working on ranches in eastern Montana before serving in the Mexican War.

In 1930 his father-in-law offered a 400-acre ranch on share crop basis to John and his wife, Ruth. In July, 1932, John bought 69 cows and 2 bulls from the Northern Land Company. They originally came from Frances K. Holdrege, Chadron, Nebraska, whose herd dated to the early 1920's. One of the bulls selected was

Polled Toonie, shown at the 1930 Des Moines, Iowa, National. The cattle were acquired on contract, with no money down, to be paid for at $125 per head, credited on the choice of the bulls taken each fall from the calves produced. Rice bought MP Domino 2d, four years old, in February, 1935, delivered to Montana for $200. Ruth states that MP Domino 2d was one of Rice's greatest finds. MP Domino's greatest son was Plato Domino 36th. Trumode Domino 8th, calved October, 1942, and judged 1944 National Champion bull at Atlanta, Georgia, was perhaps the greatest son of Plato Domino 36th.

Rice's first love was Polled Herefords. His life ended over Hawaii, when he was flying to see Parker Ranch Polled Herefords. The plane that carried him was lost April 17, 1954.

In less than twenty-two years he had built up a great herd of cattle, served three sessions in the Montana legislature, been awarded an honorary colonelship by the governor of Montana, purchased two large estates, great coal interests, and Lake DeSmet—all on a "shoestring," as he would tell you. He did all this with Polled Herefords—a Horatio Alger story?

Rice was posthumously inducted into the Hall of Fame in 1967.

Mans Hoggett

Mansfield "Mans" Hoggett was born in Gonzales County, Texas, on September 20, 1889. Mans became a towering figure in the Polled Hereford industry during the first half-century of the breed's existence. In partnership with R. A. Halbert from 1927 to 1940 and operating as the Halbert & Hoggett Ranch firm that made the HHR prefix world-famous, Mans was instrumental in developing the Domestic Mischief and Domestic Anxiety families so popular today in both the United States and Canada.

Hoggett was elected to the board of the APHA in 1939, served as president in 1947, was re-elected to the board two years later, and served until 1953. He also presided over the Texas Polled Hereford

Association and was on the board for eight years. He served as president of both the Brown County and the Concho County Polled Hereford associations.

His leadership and enthusiasm for the breed was a source of encouragement to many others during difficult times, especially during the depression of the 1930's and the periods of threat to the fledgling breed of Polled Herefords. His lifelong companion, and an ardent promoter of Polled Herefords was his wife, Marie.

Hoggett was inducted into the Hall of Fame in 1967.

M. P. "Hot" Moore

M. P. "Hot" Moore, the man who made CMR almost a byword in every nation where Polled Herefords are bred, was the tenth man inducted into the Polled Hereford Hall of Fame.

Moore—whose nickname "Hot" has stuck since he pitched baseball for the University of Alabama (and lost only one game in three years)—was born in 1905 in Senatobia, Mississippi. In 1933 he started a registered Polled Hereford herd with purchases from the White, Harvey, Trenfield and Halbert herds in Texas; the Lewis herd in Kansas; and the Blanchard and Kuhlmann herds in Nebraska. He bought his first herd sire, Pawnee Rollo 44, from Hugh H. White, Keller, Texas. From this start and Moore's uncanny ability to select and linebreed sprang the CMR Rollo Domino and CMR Advance Domino families, foundation for the famed CMR herd.

Moore served three times as president of the American Polled Hereford Association, chaired the building committee for the new APHA headquarters in Kansas City, Missouri, was instrumental in establishing the National as a "roving" event in 1941, and was the first Polled Hereford breeder to patent his prefix. But he is perhaps best known for his unbroken series of record-setting production sales—thirty-four of them—that have established him as a merchandiser without peer in the livestock industry. Always

Otto A. Maul

R. W. Jones, Jr.

R. E. "Pat" Connolly

Sam Swann

by his side and as knowledgeable as any cattleman, was his wife, Annie Louise.

Moore was inducted into the Hall of Fame in 1969.

Otto A. Maul

Otto A. Maul, inducted into the Hall of Fame at age sixty-seven, began his Polled Hereford operation in 1938 and has had an average of sixty-five to seventy-five cows over the years. For years Otto A. Maul & Sons' entries at the National Western Show were among the few representatives of the Polled Hereford breed in that important show, and he has been a continuous exhibitor in Denver's carlot division for many years.

A former APHA director, Maul and his wife, Lillian, have long been active in cattle associations in the Colorado-Wyoming area. He has served as president and director of Western Polled Herefords, Inc., a member of the Colorado Beef Board, a director of the National Western Stock Show, a member of American National Cattlemen's Association (ANCA) committees on beef promotion and brand and theft, a director of the Colorado-Wyoming Polled Hereford Association, and a member of the Colorado Cattlemen's Association (serving as president in 1960). He has been long active in the Colorado Hereford Association and the Elbert County Livestock Association.

Cattle from the Maul ranch were among the ten thousand head of Polled Herefords exported to Chile during the massive cattle airlifts of 1968 and 1969.

Maul was inducted into the Hall of Fame in 1971.

R. W. Jones, Jr.

R. W. Jones, Jr., purchased his first purebred Polled Herefords in 1944 and had one of the first two herds to participate in the Georgia Beef Cattle Improvement Association (BCIA) when it started in 1956. He was perhaps the Polled Hereford industry's leading proponent of performance testing during the long years when performance standards were not popular as measurements of the value of breeding cattle. In his breeding program (which was continued by his widow as R. W. Jones, Jr., Farms) he preserved a bloodline that in recent years has added immeasurably to the value of Polled Herefords as a breed. Animals bearing Jones's RWJ prefix have been used in experiment-station herds in all parts of the country, and many of his bulls have been selected by artificial insemination (AI) services.

Jones held membership in the APHA, the American Hereford Association, the Performance Registry International, the Georgia Beef Cattle Improvement Association, the National and Georgia Cattlemen's associations, and the Georgia Livestock Association. His herd still participates actively in the Guide Lines Program. Although he was not necessarily a proponent of the show ring, his cattle and their descendants have been some of the most successful in show-ring competition around the world.

His herd was dispersed July 19, 1975, for an all-time world's record price for any breed of $9,844 per lot. The herd consisted of 120 head.

Jones was inducted into the Hall of Fame in 1971.

R. E. "Pat" Connolly

R. E. "Pat" Connolly, a leading Polled Hereford breeder from California, went into the Polled Hereford business in 1953 and served as president of the APHA in 1967. He showed outstanding leadership in a time of stress for the breed and for the association. His dedication to the breed has been shown in his giving of his time to travel all over the world on behalf of the association.

He had an outstanding herd-breeding program and was the breeder of CPH Beau Mischief 3d, grand-champion bull at the 1969 National Polled Hereford Show and Sale in Atlanta, Georgia. He is a past president of the California Polled Hereford Associa-

tion and served on the Liaison Committee in contract deliberations with AHA. He was a guest delegate to the World Hereford
Conference in Sydney, Australia, in 1968. He and his wife, Doris,
attended all Polled Hereford functions throughout the United States
and around the world for many years. His contribution will long
be remembered as a lasting benefit to the breed.

Connolly was inducted into the Hall of Fame in 1971.

Sam Swann

Sam Swann, of Merkel, Texas, was the fifteenth man inducted
into the APHA's Hall of Fame. Swann figures that he has been
breeding Polled Herefords longer than any other living man. His
father was ranching near Merkel at the turn of the century, and
young Sam helped drive cattle to the rails in Merkel and Trent for
shipment to St. Louis or New Orleans, for those were the years
before the Fort Worth Stockyards were built. He got his first Polled
Hereford bulls in 1905 and about thirty years later went into the
registered business.

He is as progressive and enthusiastic about the future of the
breed as men a quarter his age. Sam Swann takes care of 150
registered cows on 2,960 acres with the help of only one man. He
rides every day and participates actively in programs of the APHA
(he has been on the Guide Lines Program since its beginning), as
well as the Texas, New Mexico, Brown County, and West Texas
Polled Hereford associations. He has served on the boards of the
four latter groups, as well as on the board of the West Texas Fair
Association and as president of the West Texas PHA. He provided
the basis for such herds as Ogeechee Farm, Fairland, Oklahoma,
and has served as friendly consultant and confidant to many.

Swann was inducted into the Hall of Fame in 1972.

Earl Purdy

Hubert M. "Pop" Mullendore

W. R. Gollihar

Dwight L. Moody

Earl Purdy

Earl Purdy was born in 1892 in Harris, Missouri, and was active in his family's Shorthorn herd operation there until its dispersal in 1929. He subsequently managed the famous Hereford herds at Milky Way Farms, Banning-Lewis Ranches, and Hillandale Farm. He later managed the Shorthorn operation at Whitecroft Farm and then went to Elcona Farm.

His association with Polled Herefords began in 1944, when he went to work as manager for E. E. Moore at Double E Ranch, Senatobia, Mississippi, and during the next eleven years he guided the herd to national prominence. One of his first acquisitions for Double E was the immortal Victor Domino 126, and it was Purdy's keen eye that early caught the potential of "the 126th's" great breeding son, EER Victor Domino 12, "Popeye."

Purdy left Double E in 1955 to manage the extensive Hull-Dobbs Ranches' operation in Mississippi and Texas, later went into herd consultant work, and then managed Spring Lake Ranch at Tupelo, Mississippi. He was manager-consultant at Caldwell Hereford Ranch, Hazlehurst, Mississippi, until 1965.

Purdy selected, fitted, and exhibited several national champion females. His outstanding contribution to the breed was recognized in the two great cow herds he developed at Double E Ranch and Hull-Dobbs Ranches. He was also recognized, however, for the many outstanding young cattlemen whom he trained at these two operations. Of the many future herdsmen and managers that developed under his watchful eye, three were his own sons, who have carried on his practical philosophy. They are: Herman, Konrad and Bruce.

Purdy was inducted into the Hall of Fame in 1972.

Hubert M. "Pop" Mullendore

Hubert M. "Pop" Mullendore, the still-active patriarch at Mullendore Hereford Farm, south of Franklin, Indiana, has roots in the purebred industry extending back to 1889, when his grandfather gave his father a registered horned heifer to launch the family's thriving Hereford operation.

To young Mullendore fell the task of showing his father's cattle at the various county fairs. On one such trip he visited the farm of J. E. Green (president of the APHA from 1909 to 1916) at Muncie, Indiana, saw the bull King Jewel 10, and bought him. After a few sessions with his father, young Mullendore was allowed to use the Polled bull on two cows and subsequently traded him back to Green for Polled Peach 10, who sired a heifer the Mullendores sold for $1,000. That heifer not only justified the young man's enthusiasm for the new hornless breed but convinced his father too.

A long-time participant in state and area breed activities and a past director of APHA, Mullendore through the years has been known for breeding useful, prepotent, middle-of-the-road cattle unaffected by fads and fancies—the kind that have contributed immeasurably to the foundation of herds across the nation and in several foreign countries.

Mullendore was inducted into the Hall of Fame in 1973.

W. R. Gollihar

W. R. Gollihar, Whitney, Texas, was the seventeenth man inducted into the Hall of Fame, and probably no other is so closely identified with one bull as he is with Mesa Domino. W. R. Gollihar, Gollihar Hereford Ranch, Mesa Domino, and Mesa Domino cattle—mention of one calls to mind the others—it was this great Texas cowman who above all established the gene pool from which the Mesa Domino family spread across the continent.

Gollihar bought his first registered Polled Herefords in 1927 from Burleson & Johns, Whitney, Texas, and soon established a breeding program that earned his cattle a wide reputation for uniformity and prepotency. He built his herd with groups of females uniform in type and breeding—usually half-sisters—and consistently sought improvement through herd-bull selections. If a new bull did not sire calves that were better than Gollihar already was getting, he went to market regardless of price, breeding, or conformation. And through the years Gollihar allowed no one to "top his herd."

Thus the foundation had been firmly established when, in the fall of 1950, Gollihar attended Harvey Lobdell's dispersal sale in Colorado and returned to Texas with the Lot 1 bull Mesa Domino, a seven-year-old bred by Hall of Fame member F. L. Robinson, Kearney, Nebraska. Bred to daughters of the horned bull Advance Return, Mesa Domino proved to be the "nick" from which Gollihar never swerved, a bull whose descendants reflect the wisdom of steadfast adherence to the goals of a breeding program still being carried on today by his family.

Gollihar was posthumously inducted into the Hall of Fame in 1973.

Dwight L. Moody

Dwight L. Moody was born on October 16, 1898, in Climax Springs, Missouri, and served with the United States Marines during World War I. He went to work for Ford Motor Company in 1919, following his discharge from the service.

Moody started in the cattle business in 1948, when he purchased an Angus herd, but he soon switched to Polled Herefords because he said, "I liked the disposition of Polled Herefords much better and I definitely thought they were the coming breed. They also had more size and substance." His first whitefaces came from Allen Engler's herd at Pauline, Kansas, in 1952.

In 1966 Moody sold his Ford dealership in order to devote full attention to his herd. He registers more than 100 calves annually from the 135-cow Lamplighter herd, maintained on his four-hundred-acre farm on the outskirts of Lee's Summit, Missouri, though the herd once numbered as high as 265 head. He has held annual production sales for more than two decades and has shown extensively, especially in association-sponsored events throughout Missouri.

Past president of the Missouri Polled Hereford Association, Moody served one term on the APHA Board of Directors, during which time he was secretary-treasurer of the committee in charge of building the present APHA headquarters offices in Kansas City.

Dwight Moody and his wife, Mabel, who were married on Christmas Eve, 1922, have two children, Dwight E. "Bud" Moody and Mrs. Roger (Letty Belle) Ayres, both of whom have been active in the family's Polled Hereford operation, and two grandchildren.

Moody was inducted into the Hall of Fame in 1974.

Samuel R. "Mr. Sam" Morrison

Jim Gill

A. G. Rolfe

John H. Royer, Jr.

Samuel R. "Mr. Sam" Morrison

Samuel R. "Mr. Sam" Morrison, was born in 1893 in Orford, New Hampshire, and was educated at the University of New Hampshire. For seventeen years he managed Brookshire Farms, a Hereford operation near Windsor, Massachusetts, and then moved to Mississippi as a feature writer for various livestock and breed publications.

During his travels as a journalist Morrison became enthusiastic about the South's potential for beef-cattle production, and in 1937 he helped organize the famous Panola-Tate Livestock Association (AAL). He served as its first and only manager and also at intervals as president, vice-president, and director during the more than thirty years the Panola-Tate group was active.

Though the Panola-Tate Association's primary objective was to promote all classes of livestock, Morrison and his contemporaries nevertheless achieved world-wide fame in the production and marketing of registered Polled Herefords. Two or more consignment sales were held annually, and cattle consigned by members were shipped into forty-eight states, as well as to eight foreign nations. Records set in the Panola-Tate auction ring made breed history for three decades.

"Mr. Sam" and his wife, Elizabeth, also operated Onandaga Ranch, just south of Senatobia, Mississippi, and sold cattle from his fifty-eight acre "1-mule operation" (as he humbly described it) to buyers throughout the country.

In 1964 "Mr. Sam" was honored by the Senatobia Chamber of Commerce for his contributions to his community and to the beef industry. Through his efforts and those of others like him, Mississippi grew from a one-crop "cotton economy" to a leader among the beef-producing states. Morrison was also a director of the Livestock Committee of the Delta Council.

Morrison was inducted posthumously into the Hall of Fame in 1974.

Jim Gill

He started in the Polled Hereford business in the early 1930's with a Mossy Plato bull and a few females purchased with borrowed money. He began concentrating on Domestic Mischiefs after producing the immortal Domestic Mischief 97, the bull whose descendants put Coleman, Texas, "on the map" and made the JFG prefix known world wide.

The JFG prefix symbolized the team of Jim and his wife, Fay, who was his constant traveling companion and attended every national show since 1939 . Their time dates back to the time and experience of the Gammons.

Among many great herd bulls produced by JFG was Diamond, long used in the Orvil Kuhlmann herd in North Platte, Nebraska. Diamond's progeny was shown extensively in open Hereford competition and was one of the highest-ranking Register of Merit sires.

Long a leader in state and area associations, Jim served on the APHA Board of Directors from 1944 to 1949, was elected again for four-year terms beginning in 1955, 1963, and 1974, and served as president in 1945 and 1965.

Jim Gill was the eighteenth inducted into the Hall of Fame. Gill was inducted into the Hall of Fame in 1974.

A. G. Rolfe

A. G. Rolfe was the nineteenth man inducted into the Hall of Fame. An Illinois native, Rolfe moved east with his highway-paving business and maintained both horned and Polled Hereford herds but phased out the horned cattle after buying the 1947 National Champion ALF Choice Domino 6, one of the breed's most influential bulls. "The 6th" headed Rolfe's herd for almost twenty years, drawing buyers from every state and six foreign countries.

Rolfe was elected to the APHA Board of Directors in 1945,

served as president in 1946, and was re-elected to four-year terms in 1948 and 1956. He was serving as a national director when he died in 1959 at the age of seventy-seven.

Rolfe was owner and founder of Spring Valley Polled Herefords, Poolesville, Maryland where many champions were bred and shown. Among the many great bulls produced there was SV Beau Perfection and the great cow SV Benefactress 52 which became one of the breed's outstanding brood matrons at Circle M Ranch, Senatobia, Mississippi.

Rolfe was posthumously inducted into the Hall of Fame in 1974.

John H. Royer, Jr.

John H. Royer, Jr., was born in Okarche, Oklahoma, in 1906. He owned and operated the 600-acre Bushy Park Farm in Maryland from 1946 to 1967 and then a 1,050-acre farm in Virginia. The sole interest and occupation of John and his wife, Mary, since 1946 has been Polled Herefords except for three years, 1963 to 1966, when John was in government service.

He is the breeder of the BPF Pawnee Mixer family, which was one of the leading bloodlines for several years and is found in the prominent lines of today. He exported BPF Pawnee Perfect to Oscar Colburn in England in 1957. He has exported bulls to England, Argentina, New Zealand, and Canada. The sons and grandsons of BPF Pawnee Druid and BPF Pawnee Mixer are serving prominently in many countries.

Royer was a director of the APHA from 1952 to 1955 and was president in 1955. He was president of the Eastern Polled Hereford Association in 1973 and 1974. He managed the Eastern Polled Hereford Association sales for twelve years; those sales were always among the top three sales in the country.

Royer was inducted into the Hall of Fame in 1975.

Leon Falk, Jr.

Dick Hibberd

R. L. Swearingen, Sr.

C. E. "Ernie" Knowlton

Leon Falk, Jr.

Leon Falk, Jr., of Pittsburgh and Schellsburg, Pennsylvania, has left an indelible mark on the breed. Formerly in the registered Jersey business, he converted his farm to a Polled Hereford operation in 1951. He has produced and shown many champions, including the national champion bull in 1969 and 1973. He has established many new breeders in the business and has helped develop improvement of the breed. He has unselfishly funded research projects; one in particular was the five-year horn-scur research program at Pennsylvania State University. He was a national director of the APHA from 1967 to 1971. While serving as president, he influenced the change of the corporate structure of the association to a more modern pattern and helped initiate the American Beef Records Association (ABRA) program to allow the APHA to record for other breeds. He was an active leader in the Eastern Polled Hereford Association and encouraged establishment of many new herds in the eastern United States. His influence has had a profound effect on the breed. The continued interest on the part of Leon and his wife, Loti, in the Polled Hereford breed is a source of encouragement to many.

Besides his Polled Hereford interests Falk has been a successful and prominent industrialist with substantial interest and leadership in the steel industry. The Falk Foundation, Pittsburgh, Pennsylvania, of which he has been chairman, has been dedicated to many humanitarian projects. Few men have had such wide range of interests and asserted as much progressive influence.

Falk was inducted into the Hall of Fame in 1975.

Dick Hibberd

Dick Hibberd and his wife, Laurose, of Imbler, Oregon, established Hibberd's Hornless Herefords in 1943. He has been a leader of Polled Hereford breeders in the Northwest since his entry into the Polled Hereford business. He has not only supplied the commercial cowman with seed stock but has helped many new breeders establish Polled Hereford herds. His quick wit, sense of humor, and ready smile have been his trademarks at the many functions he has attended. He was a director of the APHA from 1953 to 1957.

Hibberd's beginning with registered Polled Herefords was in 1932, when he acquired fourteen cows and one bull at a bank foreclosure sale. Four of the cows he purchased came originally from the Gammon farm at St. Marys, Iowa, the birthplace of Polled Herefords. His first bull was Mossy D Bullion-109452. King Woodford 2-122818, bred by Dingwall Company, Drummond, Montana, was reported to be his most outstanding herd sire and was known as a fountainhead sire of the Northwest. King Woodford was a son of King Domino.

Hibberd was a prime leader in organizing the Columbia Empire Polled Hereford Association and served as its president for two years. He also served as president of the once-active National Western Polled Hereford Association.

Hibberd was inducted into the Hall of Fame in 1975.

R. L. Swearingen, Sr.

R. L. Swearingen, Sr., and his wife, Linda, began breeding Polled Herefords in 1947 and became a member of the association in 1950. He has recorded up to 128 head of Polled Herefords a year and has traveled many miles to promote Polled Herefords throughout the Southeast and elsewhere. He has supported sales throughout the country and was elected to the Board of Directors of the APHA in 1960. He was president of the APHA in 1964 and was an observer representing Polled Hereford breeders at the 1964 World Hereford Council in Dublin, Ireland. It was at that time and through his efforts that the APHA became a full member of the World Hereford Council and the APHA's registration certificates were recognized world wide for export.

Swearingen remains active in the Polled Hereford business. The Standard of Perfection Show in Atlanta, Georgia, in October, 1974, was held in his honor. Not only has he been an active and dedicated Polled Hereford breeder, but he has been a leader in his community at Reynolds, Georgia, in many civic and benevolent activities. His leadership ability has been reflected from the local to the national level, and he has served voluntarily wherever he saw an opportunity to promote progress.

Swearingen was inducted into the Hall of Fame in 1975.

C. E. "Ernie" Knowlton

C. E. "Ernie" Knowlton, of Bellefontaine, Ohio, began his herd and joined the APHA in 1950. Ernie and Caroline Knowlton were familiar faces and ardent promoters of the breed for nearly twenty years. He set new records in the show ring, having several national champions and, on one occasion, in 1958, he exhibited both the National Champion Bull and Female, among other winnings. He also had Champion Sale Bull.

Ernie was one of the most influential people in the planning and construction of the APHA building in 1958. He gave generously of his time, talent, and money.

National Champions exhibited by CEK (Knowlton's prefix) were CEK Zato Mischief, CEK Mixer Return, CEK Zato Tonette 1, CEK Royal Lady 12, and CEK Lady Return 69.

Knowlton conducted annual production sales from 1960 to the herd's dispersal in 1973. He was instrumental in establishing many new herds through Ohio and the East. He provided outstanding leadership for the Buckeye Polled Hereford Association, serving as its president for several terms.

Knowlton was inducted into the Hall of Fame in 1975.

J. E. "Ernest" Lambert

Neil Trask

Paul V. Pattridge

Colonel E. Brooke Lee

J. E. "Ernest" Lambert

R. E. Lambert & Sons was a prominent Polled Hereford operation for many years at Darlington, Alabama. Ernest, a son, was one of the staunchest promoters of the breed until the dispersal of his herd about 1962. He was president of the APHA in 1943 and again in 1956 and was one of the few who committed his herd to complete single recording with the APHA. Few men had as much to lose as did Ernest Lambert. He followed through with his commitment and maintained a single APHA recorded herd. Unfortunately, the entire breedership did not follow his lead, and when he dispersed, he took a severe economic loss as the result. He contributed greatly through his own personal sacrifice toward lifting the double-registration yoke from the necks of Polled Hereford breeders. He died a short time after the dispersal.

Through his leadership, determination, and encouragement he was a great influence on the Polled Hereford breed in the Southeast and established new standards for breeders to follow throughout the United States.

Lambert was posthumously inducted into the Hall of Fame in 1975.

Neil Trask

Neil Trask, of Calhoun Falls, South Carolina, began breeding Polled Herefords in 1933 and became a member of the APHA in 1937. He has recorded some Polled Herefords each year and recorded his highest number in 1968, 282 head. Trask had a fountainhead herd in South Carolina for many years, breeding primarily Plato Domino cattle. He perpetuated the Plato Domino family of cattle which has been the base of many cow herds throughout the United States. He never believed in "pampered" cattle and gained a reputation for producing hard-working cattle-men's kind of cows with mothering ability. He has had as a primary goal "to produce cattle that would grow fast, fleshen, and thicken on grass."

Trask has supplied the cattle for many new herds and the seed stock for the improvement in many established herds. He is still active in the Polled Hereford business.

Some of the great bulls produced by Trask were NT Rupert, P9446404; Victor Plato 35, P71314476, and Plato Domino 43, 3080818. He has recorded more than four thousand Polled Herefords in his breeding career. Trask's cow herd has long been recognized as one of the outstanding producing herds, and his replacement heifers have been greatly sought after by old and new breeders alike.

Trask was inducted into the Hall of Fame in 1975.

Paul V. Pattridge

Paul V. Pattridge, Golden, Colorado, started in the Polled Hereford business as Johnson and Pattridge in 1935. He bought his first Polled Herefords from Harvey Lobdell, Gunnison, Colorado. He recorded 50 to 75 animals each year until he sold his herd to Pioneer Cattle Company in 1973. His herd has numbered as many as 350 head at its peak. His greatest herd bull was Pat Goldmine, a product of his own breeding program.

Pattridge and his wife, Flo, have been strong supporters and outspoken advocates of Polled Herefords. He has lived and promoted Polled Herefords in a traditional horned Hereford stronghold but has been a mainstay in the polled promotional effort. He has served twice on the APHA board and was active in encouraging the Rocky Mountain area to remain loyal to Polled Herefords. He is a director in the National Western Show and has been a strong factor in getting polled shows established there. He was one of the early participants on the APHA's Guide Lines Program and adhered to it religiously, using the records to improve the performance of his herd. He also introduced English breeding into his herd with great success.

Among his other interests have been banking, where he served as cashier and finally president of the First National Bank of Golden, Colorado. He also served as Jefferson County assessor and county treasurer.

Pattridge was inducted into the Hall of Fame in 1975.

Colonel E. Brooke Lee

Colonel E. Brooke Lee, with his wife, Nina, Damascus, Maryland, started in the Polled Hereford business as a member of the American Polled Hereford Association in 1955. For three years in a row he recorded more Polled Herefords than any other breeder in the world. The colonel has been one of the breed's most colorful characters, as well as a success in many businesses. His innovative, promotional efforts and strong personal appeal have resulted in putting more people in the Polled Hereford business in the East than probably any other breeder. His contribution in support and consignments have contributed greatly to the success of Polled Herefords offered in public auction in the eastern part of the United States. At eighty years of age he is still active and talking of expanding his herd. He has operated Polled Hereford farms not only in Maryland but also in Mississippi and Missouri. The first Polled Hereford show in the state of Virginia was held in his honor in Richmond in 1974.

Lee's zeal, enthusiasm, and leadership were reflected in his activities for many years in the Eastern Polled Hereford Association and subsequent expansion to Mississippi and Missouri.

Lee was inducted into the Hall of Fame in 1975.

Otha H. Grimes

Otha H. Grimes, of Tulsa and Fairland, Oklahoma, is one of the breed's most progressive breeders. He established his herd in 1951

Otha H. Grimes

Ken Davies

Henry Kuhlmann, Jr.

and was one of the crusaders in the early performance-testing days. Although he is nearing eighty years of age, he maintains his herd of nearly five hundred cows and is actively engaged in promotion world wide. He has exported a great number of cattle and has been instrumental in influencing many breeders to keep records. His annual production sale of Guide Lines cattle has contributed greatly to the improvement of the breed.

Grimes and his wife, Mildred, established the famed Ogeechee Polled Hereford Ranch on Domestic Anxiety and Domestic Mischief bloodlines. He was not active in show-ring activities with his herd of cattle for many years, although he produced many cattle that performed well there for others. His exhibition was confined mainly to carload lots at Denver, where he was very successful, showing several champions. He and his well-known manager, Odell Gelvin, were a working team without peer. They were successful in exporting cattle throughout the world because of the herd's famed reputation as high-performing cattle.

Grimes was inducted into the Hall of Fame in 1975.

Ken Davies

Ken Davies of Cholame, California, became a member of the APHA in 1941. He was owner and operator of the X-D Cattle Ranch, and made his brandname famous world wide. He bred many outstanding Polled Herefords and was instrumental in establishing a great number of new breeders in the business. He served as the long-time secretary-manager of the California Polled Hereford Association and was a key factor in establishing the famed Poll-O-Rama at Sacramento, California. He and his wife, Lilah, were familiar faces at all Polled Hereford functions on the West Coast and gave generously of their time and selves to the Polled Hereford effort. He served as a director of the APHA from 1955 to 1959. His greatest herd bull was Battle Domino 7. He dispersed his herd in 1970.

Ken Davies was a seed-stock producer without peer for many years on the West Coast. He gave great emphasis to cow-herd efficiency and reproduction in his breeding and selection program. He provided the basis for beginning herds in many areas, including Japan and Hawaii.

The Polled Hereford breed was given great impetus, and much progress resulting from the tireless effort, fine leadership, and influence of Ken Davies.

Davies was inducted into the Hall of Fame in 1975.

Henry Kuhlmann, Jr.

Henry Kuhlmann, Jr., was born on February 18, 1882, in Champaign County, Illinois. When he was five, his parents moved to southeastern Nebraska, where they raised commercial cattle until 1896, when Henry Kuhlmann, Sr., began his registered Hereford herd. Henry, Jr., had six brothers and an adopted sister. Henry received the best of livestock training from his father.

It was in 1917 that Henry Kuhlmann, Jr. bought his first Polled Herefords at a sale in Manhattan, Kansas. He bought seven heifers, five of which he kept to form the nucleus of his future famous herd and which were the basis of all the Kuhlmann families' foundation breeding.

His first matings were with a horned bull named Daylight. His second bull was a polled bull, Wonderful 6th by Wonderful by Bullion 4th. Following Wonderful 6th was Beau Blanchard 50th. Kuhlmann stated in an interview in 1951:

"One reason I liked Polled Herefords went back perhaps to my boyhood days on the farm. One of my early jobs was dehorning cattle both for my father and for neighbors, and I never liked it. And I'll say here that I ran across quite a few horns on a breed that was not supposed to have had any from the start. When I heard about Polled Herefords, I became interested, maybe because of that job I'd had."

Kuhlmann's most famous herd bull was Choice Domino, an October 2, 1923, son of Prince Domino out of Dorana 20th by Beau President. He later purchased Advance Domino 47th and bred him to daughters of Choice Domino. It was from one of these matings to Linda Domino that resulted in Advance Domino, the sire of Advance Domino 30 and Advance Domino 50, foundation sires in two of the herds of Kuhlmann's sons, Orvil and Kenneth.

Kuhlmann was inducted into the Hall of Fame in 1975.

The Hall of Merit

The American Polled Hereford Association's Hall of Merit was established in 1971. It serves as a very appropriate way to grant recognition and appreciation to individuals that serve the beef cattle industry in an unselfish and dedicated manner. The Hall of Merit inductees' activities must not necessarily be confined to the Polled Hereford industry. The honor is usually conferred upon those who have made a contribution to the industry in a special field. The area designations are: Education & Research; Communications; Public Affairs; and Youth. Qualifications for installation in the Hall of Merit in special fields are as follows:

Education & Research

This award may be received by any person employed in the field of beef cattle research or related fields and those distinguishing themselves in the field of education on the college or high school level. Only those persons who have dedicated time and effort and exhibited special interest on a national scope are eligible to receive this award.

Communications

Those persons employed in the field of publications, public

relations, promotion or related activities may be eligible for this award. One must have displayed an unusual interest in advancing the common good of agriculture and a special personal effort and concern for the public image of the purebred beef cattle industry.

Public Affairs

Any person active in the field of public affairs and demonstrating dedicated effort toward bettering relations between the government, the public and agriculture is eligible for nomination. One must have, through personal interest and concern and special effort, enhanced the image and economic status of the beef cattle industry to receive the award.

Youth

This award may be given each year to the person recognized as having made long, continuous, outstanding contributions to the advancement of youth in the area of Polled Hereford activities. One must have through personal effort in encouragement, leadership or monetary contribution, enhanced the youth program on a wide scope, and one must have shown a special personal interest in youth and through his efforts made a lasting contribution in encouragement and leadership to inspire youth to higher goals.

Listed here are the names of those who have been elected to the Hall of Merit through 1975, with the positions in which they served and made their contribution.

Education & Research

L. S. "Bill" Pope, Ph.D., Assoc. Dean of Agriculture, Texas A&M University, College Station, Tex.

Bryon L. Southwell, Georgia Coastal Plains Experiment Station, Tifton, Ga.

Keith Gregory, Ph.D., Director, USDA, Meat Animal Research Center, Clay Center, Neb.

Lowell Wilson, Ph.D., Professor of Animal Science, Pennsylvania State University, University Park, Pa.

Frank Baker, Ph.D., Dean and Director, College of Agriculture, Oklahoma State University, Stillwater, Okla.

Jim Wickersham, General Manager, Triangle T Ranches, Oakland, Cal.

Communication

Johnny Jenkins, Editor-Publisher, *Livestock Breeders Journal*, Macon, Ga.

Frank Farley, Sr., Publisher-Editor, *Polled Hereford World*, Kansas City, Mo.

Frank Reeves, Livestock Editor, *Ft. Worth Star-Telegram*, Ft. Worth, Tex.

Forrest Bassford, Publisher, *Western Livestock Journal*, Denver, Colo.

Foss Palmer, Director of Advertising Field Services, *Chicago Drovers Journal*, Chicago, Ill.

Roderick Turnbull, Agricultural Editor, *Kansas City Star*, Kansas City, Mo.

Public Affairs

W. D. "Bill" Farr, American National Cattlemen's Association, Greeley, Colo.

W. R. Watt, Sr., President, Southwestern Exposition & Fat Stock Show, Ft. Worth, Tex.

A. A. "Tony" Buford, Polled Hereford breeder, Caledonia, Mo.

John Trotman, American National Cattlemen's Association, Montgomery, Ala.

Hayes Walker, Sr., Editor, Publisher, *American Hereford Journal*, Kansas City, Mo.

Youth

Mrs. Mona Chisholm, California Poll-ettes President, Santa Rosa, Calif.

O. G. Daniel, Ph.D., Head, Extension Animal Science Department, University of Georgia, Athens, Ga.

Brownell Combs, Polled Hereford Breeder, Lexington, Ky.

Mrs. Helen Hartfield, National Council of Poll-ettes President, Granville, Ohio

Mrs. Frances Lewis, Larned, Kans.

The Poll-ettes

It was in the fall of 1958 in Sacramento, California, that the seed was sown that grew into the National Federation of Poll-ettes. Bob Jones, President of the California Polled Hereford Association, gathered a group of twelve women, wives of Polled Hereford breeders, and asked them to decorate the tables for the annual Poll-O-Rama banquet. It was an eager and willing group, headed by Mona Chisholm. Little did they realize at the time that from this single assignment would begin the national movement.

It was a suggestion by Mrs. Don Case that this group of women could make a positive contribution to the promotion of Polled Herefords in California if they encouraged youth by giving trophies at the fairs. The first trophy program was initiated in 1959 and since that time literally thousands of trophies have been given and hundreds of youth have been encouraged to have Polled Hereford projects that ultimately resulted in Polled Hereford herds being developed. It was from this original movement in California that the idea spawned and grew to other states throughout the country.

Breed associations and breed societies, like other organizations, grow and develop through evolutionary process similar to the early day family farm houses. They start out with one or two rooms when a young couple got married and then as children were born and added to the family, additional rooms would be built and added on to the original structure. Later, when the children grew up, got married and chose to stay and work on the home farm, there would be a shed or lean-to added to the original home and ultimately it would become a structure much larger and of different shape than the original.

The ladies' auxiliary to the American Polled Hereford Association, known as the Poll-ettes, is just such an addition to the original APHA structure.

The first meeting on a nationwide basis took place in Stillwater, Oklahoma, in the summer of 1965 when Mona Chisholm was invited to speak to a national gathering of wives of Polled Hereford breeders. It required about two years from that original meeting for the ladies to complete their organization and make preparations for what is now the National Poll-ette Federation.

Texas was the next state to organize into a Poll-ette state group, following California, and Pat Durham was elected as its first president.

Although it is a little known fact among breeders today, there was an early day predecessor to the National Poll-ette Federation and it was known as the "Annie Pickett Club." It was organized in Des Moines, Iowa, around 1918. Annie Pickett was the wife of Warren Gammon who devoted so much of her time and effort in her later years to the promotion of Polled Herefords and generating interest and enthusiasm in Polled Hereford wives.

The small organization of wives who gathered each year at the National Polled Hereford Show & Sale in Des Moines, Iowa, formed the nucleus of this organization and named it in honor of Annie Pickett Gammon.

In an account written by Mrs. B. O. Gammon which appeared in the February, 1923, edition of the *American Hereford Journal*,

Mrs. Gammon was credited with this writing:

"During the third Polled Hereford Week in 1918 a very few women accompanied their husbands to Des Moines. Having watched the growth and development of their cattle in preparation for the show and sale, these women also wanted to see them go through the ring. Though the men appeared surprised at this innovation, no objection was raised, and the number of women at this annual event has gradually increased until the attendance now ranges from 15 to 20.

"Nine women were present at the organization of the Annie Pickett Polled Hereford Club, which was named in honor of the wife of the originator of Polled Herefords, Mrs. Warren Gammon, whose maiden name was Annie Pickett. For some years the purpose of the club was largely social. It afforded a medium through which the women who were interested in Polled Herefords could become better acquainted, attend the shows and sales, discuss the merits of the cattle and become more enthusiastic boosters and more intelligent helpmates.

"When the date of the 1923 show was announced the Annie Pickett Club offered one of the silver trophies to be competed for, and at this banquet, Jan. 30, Mrs. C. H. Zybell, the president, presented this trophy to C. E. Brown, Rushville, Ill., winner of the class for the best three bulls owned by exhibitor.

"Twenty-four women were present at the annual meeting of the club this year, a substantial increase over the first meeting 5 years ago. Kansas, Colorado, Illinois and Iowa are represented in the club's membership.

"At the recent annual meeting officers for the ensuing year were elected as follows: president, Mrs. C. H. Zybell, Lake City, Ia.; vice-president, Mrs. H. L. Schooley, West Liberty, Ia.; secretary, Mrs. B. O. Gammon, Des Moines, Ia.; treasurer, Mrs. John J. Kelleher, Patterson, Ia. In view of the popularity of the club trophy at this year's show it was unanimously voted to offer another trophy for the 1924 show, and to make this trophy an annual event."

In intent and purposes, the Annie Pickett Club was very similar to the modern day National Federation of Poll-ettes. It served as a communication medium for the ladies who were interested in promoting Polled Herefords.

Today's national organization, however, is more youth-oriented in that there are a great number of opportunities today for the ladies to participate in activities related to youth. They generate enthusiasm and help give recognition to the outstanding youth who merit it.

The Poll-ettes of today have many worthwhile activities and make a tremendous contribution to the national promotion effort for Polled Herefords.

State Poll-ette organizations include Alabama, Arkansas, California, Colorado-Wyoming, Illinois, Indiana, Iowa, Kansas, Kentucky, Louisiana, Michigan, Minnesota, Mississippi, Missouri, Montana, Nebraska, North Dakota, Ohio, Oklahoma, Oregon, South Dakota, Tennessee, Texas, Washington and West Virginia.

Poll-ettes who have held the position of National Chairman of the Poll-ette Executive Planning Committee include Mrs. Jim (Fay) Gill, Coleman, Texas; Mrs. George (Fay) Ellis, Chrisman, Illinois; and Mrs. Homer (Bernadine) Fleisher, Jr., Knoxville, Illinois.

Those serving as President of the national organization have included Mrs. Noel (Helen) Hartfield, Granville, Ohio; Mrs. John (Charmaine) Szyperski, Pinconning, Michigan; and Mrs. Paul V. (Flo) Pattridge, Golden, Colorado.

14. Challenges Ahead

A Look to the Future
Motivating Changes

Challenges Ahead

A Look to the Future

How can a breed avoid the pitfalls of its predecessor breeds? How can a breed ensure its future growth and progress on a continuous, long-term basis? Is every breed or organization bound to the natural laws of metamorphosis, or can it be the captain of its destiny and steer clear of the obstacles that contribute to the slowdown and stagnation that accompanies aging and maturation? What can a breed association do to revive and renew its members and maintain the youthful zest and enthusiasm that is so necessary to growth and progress?

We might think of the breed association as John Gardner does the ever-renewing society in his book *Self Renewal* (Harper and Row) when he states:

"Every businessman knows of some firms that are 'on their toes' and others that are 'in a rut.' Every university president recognizes that some academic departments are enjoying exceptional vitality while others have gone to seed.

"What are the factors that account for such differences? It is a question that has never been examined systematically. Closer study will reveal that in all the examples given the same processes are at work. They are the processes involved in the rise and fall of human institutions. Rome falling to the barbarians, an old family

Young cattlemen with their heifer exhibits at State Fair.

firm going into bankruptcy, and a government agency quietly strangling in its own red tape have more in common than one might suppose.

"When organizations and societies are young, they are flexible, fluid, not yet paralyzed by rigid specialization and willing to try anything once. As the organization or society ages, vitality diminishes, flexibility gives way to rigidity, creativity fades and there is a loss of capacity to meet challenges from unexpected directions. Call to mind the adaptability of youth, and the way in which that adaptability diminishes with the years. Call to mind the vigor and recklessness of some new organizations and societies—our own frontier settlements, for example—and reflect on how frequently these qualities are buried under the weight of tradition and history.

"Everyone who has ever shared in the founding of an organization looks back with nostalgia on the early days of confusion and high morale, but few would really enjoy a return to that primitive level of functioning.

"A society whose maturing consists simply of acquiring more firmly established ways of doing things is headed for the graveyard—even if it learns to do these things with greater and greater skill. In the ever-renewing society what matures is a system or framework within which continuous innovation, renewal and rebirth can occur."

First we must realize that people make cattle; cattle do not make people. Every breed is the result of what people have done to it. Even those breeds that are the result of natural selection develop into types and characters because man has allowed them to. Students of genetics can bring about planned change. It might be well to recognize that in some instances cattle that have been allowed to produce in a natural and uninhibited manner have made more progress than those affected by man's attempts to plan their matings. But, generally speaking, progress has been made as

Sale day at CMR Ranch, Senatobia, Miss.

Watching the bull parade, 1972 National Western Stock Show.

the result of man's thinking and planning.

The self-renewing element of a breed association is the same as that of any other organization. It must be accountable to its members, accessible to the people it serves, and responsive to the changing needs of the industry.

Membership involvement is the key to keeping an organization vibrant and alert. An organization must make all members feel that they are a vital part of the plans and purposes. The democratic process, in which members elect their leaders and breed planners, is the best assurance of involvement and account-ability. The organization must be open and honest with its members, and the least member must have the same voice and access as the strongest or largest. The breed association must be everybody's experience. Rules must apply to all, but the fewer rules an organization has to police or enforce the better the organization can function in an atmosphere of harmony and cooperation. Accounts must be open and records accessible to all members at appropriate times. This does not mean that an indiscreet person should be able to disrupt the normal working procedure of the organization to conduct personal investigations, but simply that complete and accurate reports must be made at appropriate times. The association's leaders must have the complete confidence of its membership in their integrity and motives. There is an old Spanish proverb that goes, "Nadie lleva un jarry debajo de su abrigo con un buen proposito." Translated, it means, "No one carries a jug under his coat for an honest reason."

Too many organizations are dedicated to perpetuating the organization rather than the ideal. An important philosophy in the continued growth of an ideal is that it really is not important what it is called as long as the idea or concept lives on.

The Influence of Sound Leadership

The future of a breed is held totally in the hands and minds of men. We now have the history of other breeds and organizations to study and their mistakes from which to benefit as well as our own. We observe others now in different stages of development, some at the apex of their life span, others matured and over the hill, and still others new, enthusiastic, and hopeful.

The most common mistake made by many men and organizations is to be lulled into a false sense of security by the passing tide of time. A breed leader was heard to say, "One of the things we have going for us is we have been here a long time." All things are transitory and even at best leadership can only extend the life of a breed by using the best unbiased judgment at the most appropriate time until less discreet men gain control.

The greatest and most genetically efficient breed developed can continue to be so only if the leadership maintains the confidence of the public and its credibility with the industry. "Small men cast a long shadow in the evening." When men of short vision and selfish tendencies gain control and begin to exploit the reputation of a breed and the generations of improved genetic material accumulated and arranged by dedicated pioneer breeders through the years of selection, then the days of that breed are numbered. The democratic process of electing leadership by the people from among themselves has yet to be improved upon.

Some aspiring leaders can polish up their image and look good to their citizenry for a period of time. They may, through good public relations, work their way up the public ladder, but time and exposure will reveal their gross weaknesses and the same process that placed them in an influential position will remove them. Inasmuch as the future of the breed is in the hands and minds of the breeders, a logical, workable, and efficient organization is the key to breed success.

Leadership is elected by the breeders and given broad powers to function and establish policy based on the philosophy reflected from the grass roots breeders. Likewise, while the director-leaders have access to more information and tools in the form of professional staff, they must, in essence, assume responsibility for use of these resources to make sound policy decisions.

One of the greatest mistakes made by an organization is

Talking it over: Earl Purdy, left, manager, Double E Ranch, Senatobia, Mississippi and Sam Morrison, Panola-Tate Association manager, Senatobia, Mississippi, 1953 American Royal, Kansas City, Missouri.

investing all authority in the Board of Directors and then have the Board fail to delegate authority to competent staff personnel. The daily decisions faced by a modern organization cannot wait for a semi-annual Board of Directors' meeting. In order for an organization to function efficiently, authority must be delegated so decisions may be made expeditiously within the broad framework of policy established by a board of elected leaders.

Talent Is a Resource

From the time of policy making to the point of initiation and implementation, administration must assume the responsibility. It is the job of the Board of Directors to establish policy and evaluate the job administration does of implementing that policy when given a broad, flexible framework within which to operate. The greatest asset of a breed is the opportunities to use talented men developed in the complex business world. Breed associations have the unique opportunity of drawing from the ranks of breeders—men that have had experience in all phases of business.

The American Polled Hereford Association has had men on its Board of Directors from every profession and walk of life. Each has had something unique to contribute and has left his finger prints on the breed's programs, policy, philosophy, and image. These men—doctors, lawyers, corporate leaders, merchants, bankers, and yes, professional cattlemen—have made the breed what it is today and will assure its future. They are our greatest asset.

A modern, efficient organization has its structure clearly defined and easy to communicate. Layers of authority and complex organization breeds bureaucracy, clumsiness and slow response to the needs of individuals and the industry. Overlapping authority creates confusion. Listed on the following page is an organization chart that is workable and covers all areas of activities in which a breed association may become involved.

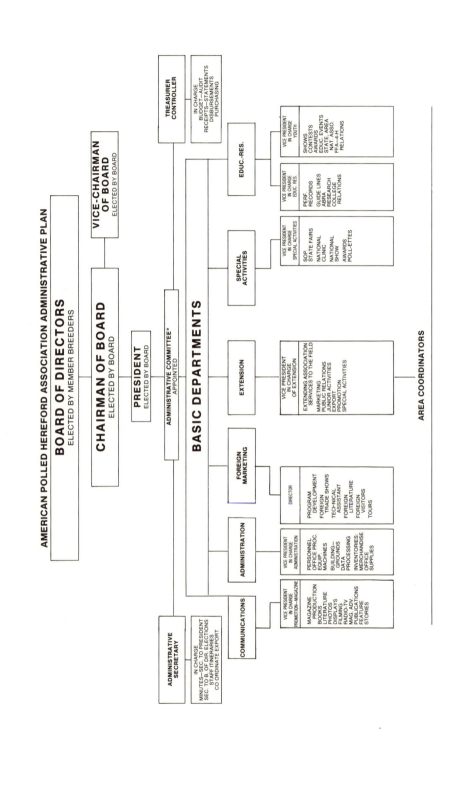

AMERICAN POLLED HEREFORD ASSOCIATION ADMINISTRATIVE PLAN

BOARD OF DIRECTORS
ELECTED BY MEMBER BREEDERS

CHAIRMAN OF BOARD
ELECTED BY BOARD

VICE-CHAIRMAN OF BOARD
ELECTED BY BOARD

PRESIDENT
ELECTED BY BOARD

ADMINISTRATIVE COMMITTEE*
APPOINTED

TREASURER CONTROLLER
IN CHARGE
BUDGET—AUDIT
RECEIPTS—STATEMENTS
DISBURSEMENTS
PURCHASING

ADMINISTRATIVE SECRETARY
IN CHARGE
MINUTES—SEC. TO PRESIDENT
SEC. TO B. OF DIR. ELECTIONS
STAFF ITINERARIES
CO-ORDINATE EXPORT

BASIC DEPARTMENTS

COMMUNICATIONS
VICE PRESIDENT
IN CHARGE
PROMOTION—MAGAZINE
MAGAZINE
PRODUCTION
BOOKS
LITERATURE
PHOTOS
DISPLAYS
FILMING
RADIO-TV
MAG. ADV.
PUBLICATIONS
FEATURE
STORIES

ADMINISTRATION
VICE PRESIDENT
IN CHARGE
ADMINISTRATION
PERSONNEL
OFFICE PROC.
EQUIP.
MACHINES
BUILDING—
GROUNDS
DATA
PROCESSING
INVENTORIES:
MERCHANDISE
OFFICE
SUPPLIES

FOREIGN MARKETING
DIRECTOR
PROGRAM
DEVELOPMENT
FOREIGN
TRADE SHOWS
TECHNICAL
ASSISTANT
FOREIGN
LITERATURE
FOREIGN
VISITORS
TOURS

EXTENSION
VICE PRESIDENT
IN CHARGE
OF EXTENSION
EXTENDING ASSOCIATION
SERVICES TO THE FIELD
MARKETING
PUBLIC RELATIONS
JUNIOR ACTIVITIES
EXPORT
PROMOTION
SPECIAL ACTIVITIES

SPECIAL ACTIVITIES
VICE PRESIDENT
IN CHARGE
SPECIAL ACTIVITIES
SOP.
STATE FAIRS
NATIONAL
CLINIC
NATIONAL
SHOW
AWARDS
POLL-ETTES

EDUC.-RES.
VICE PRESIDENT
IN CHARGE
EDUC. RES.
PERF.
RECORDS
GUIDE LINES
ABRA
RESEARCH
COLLEGE
RELATIONS

VICE PRESIDENT
IN CHARGE
YOUTH
SHOWS
CONTESTS
AWARDS
EDUC. EVENTS
STATE, AREA
NAT. ASSO.
FFA—4-H
RELATIONS

AREA COORDINATORS

Motivating Changes

The goal of a breed association should be to develop and perpetuate a breed of cattle that will economically meet the needs and desires of the consumer.

Recognizing that consumer needs are ever subject to change requires that a breed association be alert and sensitive to the need for changing standards of excellence and breed objectives. A study of the evolution of the beef industry teaches us many things, but the impressive lesson we learn is that breeds that have resisted change have become extinct or at best of secondary importance while those that have sensed changing needs and adapted to change have grown and prospered.

Just as breed associations must change and adapt to meet needs and competitive pressures so must breeders be flexible, open-minded, and resilient to industry pressures and demands. They must be extremely perceptive to detect the differences between trends resulting from real needs and fads that result from the whims of dominant or influential individuals.

It is interesting to note the descriptive terms used for beef cattle the past few decades. For many years registered breeders described cattle in vague terms and superlative phrases that were difficult to interpret into specific values. For instance, terms such as ''individual excellence, eye appeal, symmetry, and balance" were commonly used and were difficult to interpret in terms of value. The use of terms referring to specific measures such as weight, height, fertility, and reproductive efficiency were shied away from because records were not usually available.

In more recent years discriminating buyers of commercial animals and replacement seed-stock have asked for more specific information such as weight per day of age, calving interval, birth weights, and data that has a more direct influence on production efficiency. The modern cattleman is swayed less today by the esthetic value and more by the true economic values and traits that relate to volume and value of product.

The Goal—Beauty or Beef?

Breeds of the future must decide if their goal is truly economic in nature and consumer oriented, or if the aim is to simply promote interest on the part of investor-type hobbyists. The goals may be simplified by asking the question, "Is it beauty or beef for which the breed is aiming?"

Once the priority of breed values is established, then the secondary aim is to determine how attractive or eye appealing the product, or in this case, the breed, can be presented without affecting function or production efficiency.

Programs With a Purpose

A program is simply a vehicle designed to carry one person, or a group of people, from where they are to where they want to be. Viewed in this light, one can look at a breed and decide what its weaknesses are and design a plan or program to eliminate or compensate for those weaknesses. If there are deficiencies in economic traits, a program can be designed to encourage the breeder to measure his herd production and vigorously select for those traits. For example, if there is a need for increased promotion to gain a larger part of the bull market for the breed, then a program designed to motivate greater personal effort on the part of the breeder and more national promotion will be effective, provided it is wisely planned.

Program planning procedures are very important. Poorly planned and initiated programs may not be adopted and followed. On the other hand, programs properly planned and introduced may become an immediate part of a breeder's daily routine. The programs of the American Polled Hereford Association, which potentially affect all breeders—or even only a few—are thoroughly studied and researched by the staff of the APHA, reviewed by the Board of Directors, and usually approved by a committee of breeders before initiation. In some instances an advisory committee appointed by the board and the staff is called in to study the problem, and a program is developed from the committee's recommendations. Typical of the programs that were developed by the use of this procedure is the breed's Guide Lines Program for breed improvement and the Superior Sire Program for identifying superior bulls for wider use within the breed. Although these programs will not be adopted by all breeders, they will have far-reaching and profound effects upon the breed.

15. Appendix

Fiscal Year Figures
American Polled Hereford Association

Year	Registrations	Memberships Transfers	Issued	Excess of Income over Expense	Net Worth
1901	11				
1902	66				
1903	63				
1904	77				
1905	68				
1906	124				
1907	237	. . . . .	15		
1908	157	. . . . .	13		
1909	242	. . . . .	24		
1910	427	. . . . .	45		
1911	486	. . . . .	35		
1912	796	. . . . .	114		
1913	1,436	. . . . .	72		
1914	1,283	. . . . .	106		
1915	1,653	. . . . .	108		
1916	2,698	. . . . .	229		
1917	3,012	1,443	193		
1918	3,660	2,252	258		
1919	4,403	3,330	338		
1920	5,646	3,570	391		
1921	7,043	3,009	86		
1922	5,444	3,119	126		
1923	5,540	3,174	93		
1924	4,421	2,530	70		
1925	4,127	2,200	78		
1926	3,823	2,226	48		
1927	3,865	2,154	54		
1928	4,447	2,742	69		
1929	5,163	2,674	72		
1930	5,500	2,774	66		
1931	5,610	2,710	49		
1932	4,522	2,073	30		
1933	4,008	2,644	25		
1934	5,624	2,083	36		
1935	5,679	3,153	66		
1936	8,389	3,175	52		
1937	7,040	3,501	77		
1938	8,289	4,812	96		
1939	9,196	5,283	100		
1940	10,479	6,248	200		
1941	16,005	9,043	177		
1942	17,599	10,633	304		
1943	17,809	12,379	350		

Year	Registrations	Memberships Transfers	Memberships Issued	Excess of Income over Expense	Net Worth
1944	18,810	12,236	359		
1945	28,081	15,823	442		
1946	28,504	19,984	612		
1947	33,108	20,054	476		$ 44,898.68
1948	33,997	20,282	507	$ 3,854.60	48,753.28
1949	37,105	21,529	519	12,271.44	61,408.45
1950	44,263	29,927	581	22,370.42	92,993.04
1951	57,055	32,338	948	26,916.14	119,032.69
1952	71,580	37,799	1,003	48,752.61	145,908.07
1953	78,152	40,012	961	(6,586.77)	141,026.14
1954	80,020	45,823	845	(11,126.35)	125,438.05
1955	101,084	48,725	1,163	13,841.14	139,279.19
1956	106,607	58,210	2,390	17,345.12	156,624.31
1957	95,060	56,954	366	(25,828.65)	130,795.66
1958	98,328	60,115	554	35,474.72	166,270.38
1959	107,293	59,099	677	49,445.17	215,715.55
1960	111,411	59,737	634	32,216.00	309,898.65
1961	115,022	57,275	630	53,602.48	363,501.13
1962	111,088	63,189	659	24,496.56	388,089.26
1963	174,575	75,619	530	58,230.17	446,230.93
1964	175,045	89,282	443	33,778.30	479,986.53
1965	160,317	101,015	445	18,526.24	495,670.31
1966	165,062	120,011	477	18,994.78	514,665.09
1967	166,813	88,902	486	18,502.15	533,167.24
1968	132,142	77,607	3,083	(58,977.66)	664,001.03
1969	174,185	90,147	4,559	9,702.63	640,159.00
1970	159,418	83,729	3,391	44,921.43	665,427.43
1971	168,021	97,146	2,603	29,691.76	712,122.61
1972	178,391	114,369	2,659	37,610.00	596,115.00
1973	168,746	121,362	2,696	29,527.00	625,642.00
1974	207,882	113,718	2,512	45,021.00	670,663.00

Annual Sale Averages

There are few records available of public sales held in the United States prior to 1891. Since that time practically complete records are available, and they show in a graphic manner the movement of prices from year to year. The following table covers practically all the public sales of registered Herefords from 1891 to 1975 inclusive, showing the total number of lots sold in public sales each year and the average price. A lot consists of an individual animal or a female with calf at foot.

Year	Lots	Avg. Price	Year	Lots	Avg. Price	Year	Lots	Avg. Price
1891	113	68	1924	2,838	133	1955	15,251	386
1892	268	71	1925	2,805	170	1956	15,625	381
1893	61	99	1926	2,966	171	1957	15,518	386
1894	170	77	1927	3,234	204	1958	17,123	409
1895	105	91	1928	5,864	210	1959	13,461	523
1896	414	125	1929	4,198	222	1960	18,725	458
1897	205	180	1930	3,732	217	1961	14,519	471
1898	1,345	300	1931	3,564	151	1962	15,726	503
1899	1,092	258	1932	2,743	105	1963	17,454	575
1900	1,817	271	1933	4,121	94	1964	19,942	462
1901	1,895	241	1934	4,465	114	1965	21,332	453
1902	3,597	266	1935	9,681	127	1966	22,457	517
1903	2,059	172	1936	9,751	151	1967	19,356	540
1904	1,481	117	1937	9,723	193	1968	21,895	487
1905	1,179	115	1938	10,708	151	1969	24,322	503
1906	1,122	121	1939	11,120	199	1970	23,627	554
1907	1,358	124	1940	17,893	195	1971	20,104	545
1908	936	116	1941	10,048	244	1972	21,854	612
1909	1,398	127	1942	20,780	274	1973	21,876	815
1910	1,214	146	1943	25,651	353	1974	24,826	978
1911	1,203	160	1944	35,164	317	1975	27,394	689
1912	957	180	1945	36,413	369			
1913	1,707	235	1946	39,344	399			
1914	2,808	212	1947	40,411	440			
1915	3,880	233	1948	34,665	510			
1916	5,983	355	1949	36,256	540			
1917	9,157	493	1950	37,659	557			
1918	11,504	481	1951	36,087	559			
1919	19,095	491	1952	48,958	648			
1920	15,432	416	1953	68,410	926			
1921	8,516	201	1954*	63,753	865			
1922	5,086	173	1953**	12,977	550			
1923	5,235	158	1954	13,671	445			

* 1891-1954 includes totals for both horned and Polled Herefords.
** 1953-1975 includes totals for Polled Herefords only.

Trends of Major Beef Breeds

Thousand head

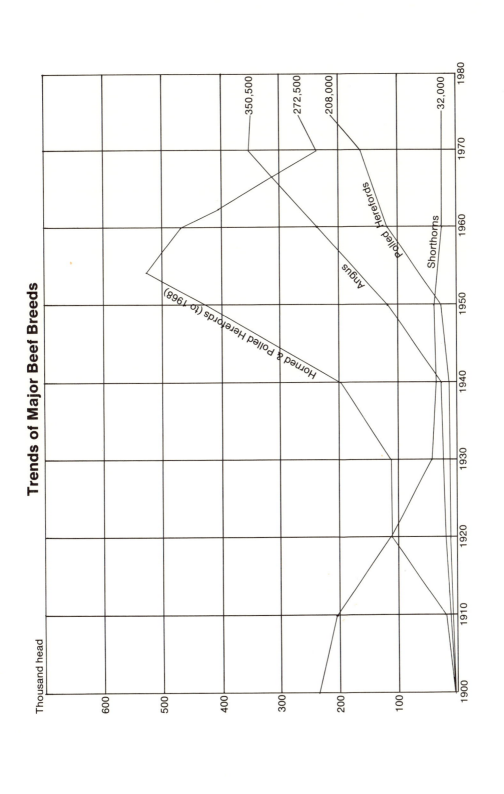

Horned & Polled Herefords (to 1968)

Angus

Polled Herefords

Shorthorns

350,500

272,500

208,000

32,000

600
500
400
300
200
100

1900   1910   1920   1930   1940   1950   1960   1970   1980

Breed History Chart of Polled Herefords

"It is not only what you know but knowing how to find what you need to know that counts."

The Polled Hereford breed history chart was designed to help the student of Polled Hereford history by providing the ancestral building blocks of the breed from the fountainhead sires to modern day family lines.

The most prominent families from their sources to some of the most noteworthy sires in the breed was the only basis until the more modern day when programs were provided to measure breeding value. The emphasis has been placed on sires because of the greater impact they have on the breed.

All bulls that have sired 500 or more progeny that were recorded in the records of the APHA are included. They are indicated with their name in parentheses and the number of progeny listed after the name. Example: (Gold Colonel) 1,806.

Most National Champion bulls and females have been included since 1928. They are indicated by having the year of their championship in parentheses. Example: Princess Dale (1928).

Sires attaining a certain level of performance on the Superior Sire program are indicated by stars.

> Superior Sire—5 Stars*****
> Senior Performance Sire—4 Stars ****
> Gold Standard Sire—3 Stars ***
> Gold Seal Sire—2 Stars **
> Gold Trophy Sire—1 Star*

Reference sires are indicated with an R in parentheses after their name. Example: HSF Sir Victor 43 (R).

Complete details of the Superior Sire program and the process for attaining the different levels of performance are available through the American Polled Hereford Association. A cursory explanation of the program can be found beginning on page 221.

MOSSY PLATO FAMILIES

```
Mossy Plato 26
    King Domino
        King Domino 31
            DK Domino
                HD Domino 22
                    HL 1 HD Dom 21
                        HL 1 Choice Domino
                            HL 1 King Dom 5919
                                HL 1K Domino 64373
                                    HL 1K Domino 68734
                        HL 1 Right Dom
                            HL1K Dom 24
    MP Domino 2
        Home Maker 2
        Plato Domino 36
            Trumode Domino 8 (1944)
                Numode 7
                    Gay Hills Misch N
                Numode 16
                    7 Up Royal Maid 13 (1950)
                    Gatesford Numode
                    HDR Royal Domino
                Numode 24
                    Advanced Numode
                Numode 29 (1948)
                Numode 110
                Numode 115
                    JR Numode J43
                        W Numaid Henrietta 2 (1960)
                Numode 116
                    EER Royal Numode 8
                GS Golden Numode 82
                    EER Royal Numode 8
            Trumode Domino 26
            Trumode 56
                My Trubaca
            Trumode 99
                JR Trumode Fast
            Trumode Domino 186
                Clayton Numode 21
                    Clayton Numode 117
                        Clayton Numode 416 (R)
```

(Continued on next page)

Mossy Plato Families (Continued)

```
Mossy Plato 26
    King Domino
    MP Domino 2
    Victor Domino
       │ Victor Domino 4
       │    │ Buster Domino
       │         Buster Domino 80
       │              Rex Domino
       │                   Rex Bocaldo 2
       │    │ CMR Rollo Domino 20 (1941)
       │         │ CMR Rollo Domino 6
       │         │    │ RCM Rollo Domino 15
       │         │         RCM Perfect Rollo
       │         │    │ RCM Rollo Domino 30
       │         │ CMR Rollo Domino 12
       │         │    │ CMR Rollotrend
       │         │         (CMR Rollotrend 5) 607
       │         │              (CMR Rollo Mixer) 521
       │         │                   CLR Rollotrend A
       │         │                        AAB Miss Roltrend 82 (1967)
       │         │    CMR Plus Rollo
       │         │    CMR Advance Rollo
       │         │    CMR Misch Dom 101
       │         │         WCH Misch R Royal 1 (1959)
       │         │    CMR Mischief Dom 5
       │         │         A Misch Domino Jr.
       │         │    GMR Advanrol
       │         │    CMR Super Rollo
       │         │         │ CMR Super Rollo 4
       │         │         │    (CMR Super Domino) 515
       │         │         │         (CMR Super Domino 147) 618
       │         │         │ CMR Super Rollo 48
       │         │         │    │ CMR Superfactor
       │         │         │    │ CMR Superol 20
       │         │         │         (SFR-CEK Superol)* 659
       │         │         │              (AAB Superol) 605 (1966)
       │         │         │    │ (CMR Superol 44) 691
       │         │         │ ORF Super Rollo 1
       │         │    CMR Mischief Dom 89
       │         │    │ CMR Blanche Dom 25 (1948-49)
       │         │ CMR Rollo Domino 28 (1945)
       │         │    │ RHR Rollo Domino 7
       │         │    │ CMR Misch Rollo 28
       │         │    │ Fox Run Rollo Dom
       │         │         Harvest King
       │         │ CMR Rollo Domino 40
       │         │    ALF Rollo Domino 23
       │         │         HVH Rollo 6
       │         │              HVH Rollo Jr. 658
       │         │ CMR Rollo Domino 47
       │         │ CMR Rollo Domino Jr.
       │ Victor Domino 2
       │    GB Victor Rollo 7
       │         PHH Prince Rollo 11
```

(Continued on next page)

Mossy Plato Families (Continued)

```
Mossy Plato 26
   King Domino
   MP Domino 2
   Victor Domino
        Victor Domino 4
        Victor Domino 126
            EER Victor Domino 2
                HSF Beau Victor 9
                    Able Victor Domino
                        HSF Vagabond Victor
                            HSF Silver Son 2
                                HSF Sir Victor 43 (R)
                HSF Beau Victor 10
                    Spidel 560
                        Spidel 445Y
                            JS Spidel 260*
                HSF Beau Victor 15
                    HSF Prince Victor 8
                        YRR Beau Prince 4
                            HSF Majestic Victor
                    HSF Prince Victor 11
                        RWJ Pr Victor F97
                            RWJ Pr Victor J133
                                RWJ Victor J133 76
                        MRF Vic Domino A14
                            (RWJ Vic Domino F18) 616
                                RWJ Victor F18 J3
                                    RWJ Victor J3 7110*
                                    RWJ Victor J3 837*
                                    RWJ Victor J3 859
                                RWJ Victor F18 K47
                                    Victorious K47 U81 (1972)
                                    Victor K47 K1
                                    Victor K47 K5
                            RWJ Vic Domino F74
                    Gay Hills Victor
                        Crutcher Victor
                            BC Supreme Victor
                        Gay Hills Victor 46
                        Gay Hills Victor 78
                        Gay Hills Victor L21
                            HR Victor Domino 33
                                HR Beau Victor 7
                        Gay Hills Victor L32
                        Gay Hills Victor L42
                            (HAB Victor Mischief) 963 (1962)
                                SVF Victor Misch 02
                                    CF Perfect Misch 87 (R)
                            SV Choice G Domino
                        Gay Hills Victor 100
            EER Victor Domino 12
                (EER Victor Tone 4) 611
                EER Victor Tone 23
                EER Victor Tone 57
                EER Victoria Tone 22 (1951)
                EER Victoria Tone 50 (1953)
            EER Victor Domino 22
            EER Victor Domino 41
            D Victoria Domino 8 (1943)
            Victor Domino Return
                Gatesford Vic Dom 11
                Gatesford Vic Dom 26
        Victor Domino 128
            Domino Lad 52
                KHF Dom Lad 13
                    Crail Victor Stanway
                        Crail Victor Dom 90**
```

(Continued on next page)

Mossy Plato Families (Continued)

```
Mossy Plato 26
  | King Domino
  | MP Domino 2
  | Victor Domino
  | Victor Domino 14
      Plato Domino 1
        | Real Plato Domino (1940)
        |   | WHF Leskan 2
        |   | Brands Sufficiency
        |   | Leskan Tone
        |         Anxiety Tone
        | Real Plato Dom 63
        | Captain Plato
        |     Captain Plato 8
        | Real Plato Dom 26
        | Real Plato Dom Jr.
        |     Colonel RPD
      Plato Domino 27
        Pawnee Domino 8
          | ALF Pawnee Mixer 21
          |   | ALF Beau Mixer 3
          |   |   | ALF Mixer Return 6
          |   |   | ALF Mixer Return 25
          |   |       Mixer Domino 25
          |   |         | D Mixer Domino 18
          |   |         |   | PVF Double Mixer 4*
          |   |         |   | PVF Double Mixer 8*
          |   |         | D Mixer Domino 36
          |   |             | Deluxe Mixer 3*
          |   |             | ALF Beau Perfect* (1967)
          |   |   | ALF Mixer Return 110
          |   |       Choice Mixer Return
          |   |   | ALF Mixer Return 115
          |   |       | (CEK Mixer Return) 605 (1958)
          |   |       |     CEK Lady Return 69 (1961)
          |   |       | CEK Pawnee Mixer
          |   | ALF Beau Mixer 7
          |   | ALF Beau Mixer 52
          |       Beau Domino Mixer
          |           Beau Domino Mixer 3
          |             JB Trojan
          |               JB Massive Trojan**
          | ALF Pawnee Mixer 24
          |   | ALF Battle Mixer 10
          |   | ALF Battle Mixer 30 (1953)
          |       | Bay Tontine 27
          |       | Bay Lady Tontine 14 (1956)
      Victor Plato 2
        | Plato Aster
        |     Plato Pioneer
        |       | Hervaleer Royal
        |       | Hervaleer Misch S
        |       | Hervaleer A
        |           | Hervalation 333
        |           | Hervaleer Misch 13
        |           | Hervaleer Perf 3
        |           | Hervaleer MP Dom
        |           | Hervalation 14
        |           | CMR Aster Domino
        |               CMR Aster Domino 7
        | Plato Aster 35
      Pure Plato Domino
```

POLLED PRINCE DOMINO FAMILIES

```
Prince Domino
  | Real Prince Domino
  |   | Real Prince Dom 24
  |   |     Real Domino 51
  |   |         Real Silver Dom 44
  |   |             Real Silver Dom 203
  |   |                 JFG Silver Mischief
  |   |                   | JFG Silver Misch 1
  |   |                   |     HAB Silver Misch 7*
  |   |                   | JFG Silver Misch 4
  |   | Real Prince Dom 66
  |   |     Real Domino 4 Sq
  |   |         TT Regent
  |   |             TT Royal Regent 1
  |   |                 LH Royal Regent
  |   |                     G Regent Misch Rollo
  |   | Real Prince Dom 75
  |   |   | Real Polled Prince 2
  |   |   |     Double Real 75
  |   |   |       | Custer Domino 9
  |   |   |       | Triple Real
  |   |   |             GS Golden Real 7
  |   |   | Real Polled Prince
  |   |       Prince Real 3
  |   |         | Prince Real 43
  |   |         |     | ACR Prince Real 12
  |   |         |     |     (Normandy Prince Gold) 647
  |   |         |     | Real Blanchard 6
  |   |         | Prince Real 46
  |   |         |     ALF Real Onward
  |   |         | Prince Real 69
  | Dandy Domino 2
  |     Colorado Dom 159
  |         MW Domino 38
  |             SLF Domino Mischief
  |                 O Larry Mischief 7 (1951)
  | Choice Domino
  |   | Choice Domino 2
  |   |     Mac Domino
  |   |         Domino Gem (1937)
  |   | Choice Domino 18
  |   |     (Choice Domino Misch) 539
  |   | Anxiety Domino
  |     | Anxiety Domino 19
  |     |     Dale Domino
  |     |         Clinton Domino
  |     | Anxiety Domino 34
  |           Pine Ridge Domino 7
  |               Pine Ridge Dom 62
  |
  |
```

(Continued on next page)

Polled Prince Domino Families (Continued)

Prince Domino
   | Real Prince Domino
   | Dandy Domino 2
   | Choice Domino
     Prince Domino Mixer
        Larry Domino
          | Polled Larry Domino
              Domino Blanchard 2
                G Larry Domino Jr.
                  | G Larry Domino 4
                  | CMR Larry Domino
                      CMR Advance R Larry
                      CMR Larry Domino 4
                          CMR Advance Larollo
                      CMR Larry Domino 21
                          CMR Larry Rollo 9
                              CMR Perfectionrol
                  | CMR Larry Domino 134
                      EJH Royal Domino 12
          Larry Gem Domino
         | Larry Domino 50
            | MW Larry Domino 36
               WAW Larry Domino
                  WW Larry Carlos
                      Miss Larry Carlos 20 (1957)
         | MW Larry Domino 96
             HHR Domino 317
     Prince Domino 1
        Colorado 21
          Colorado Domino 68
            | Clayton Domino 73
               Comprest Misch A5
                  | Comprest Domino A
                  | Comprest Domino A2
            Comprest Conqueror
               Comprest Mischief 1
           | Comprest Prince
               | Advance Comprest
                  T Mellow Comprest 3
              | Comprest Prince 2
                  Prince Comprest 26
           | Comprest Prince 25
               RBR Colorado Domino

(Continued on next page)

Polled Prince Domino Families (Continued)

```
Prince Domino
      Real Prince Domino
      Dandy Domino 2
      Choice Domino
      Prince Domino Mixer
      Prince Domino 1
      Prince Domino 4
            Prince Domino C
                  WHR Royal Domino 2
                        WHR Princeps Again
                              WHR Princeps Mixer
                                    WHR Double Princeps
                                          RS Princeps Mixer 10
                                                Mischief Princeps
                                    WHR Proud Mixer 21
                                          HG Proud Mixer 579
                                                BR Proud Mixer
                                                      JFG Domestic Mixer (1954)
                  WHR Royal Domino 51
                        OJR Royal Domino
                              Baca Domino 33
                                    Baca Duke 2
                                          Noes Baca Prince 19
                                                RHR Baca Prince
                                          RHR Baca Duke
                                                Baca Queen (1952)
      Dandy Domino
            Dandy Domino 46
                  Donald Domino
                        Donald Domino 26
                              Double Dandy Domino
                                    Double Dandymode
                                    Double Dandymode 5
                                          JR Dandymaid B12 (1954)
      Sarah Domino (1938)
```

(Continued on next page)

Polled Prince Domino Families (Continued)

```
Prince Domino
│ Real Prince Domino
│ Dandy Domino 2
│ Choice Domino
│ Prince Domino Mixer
│ Prince Domino 1
│ Prince Domino 4
│ Dandy Domino
│ Sarah Domino (1938)
│ Prince Domino Misch
│   │ Advance Domino
│   │   │ Advance Domino 47
│   │   │   Advanced Domino
│   │   │     │ Advanced Domino 30
│   │   │     │   │ Rosella (1939)
│   │   │     │   │ Stella (1941)
│   │   │     │   │ Banner Domino
│   │   │     │   │     Banner Domino 81
│   │   │     │   │         Banner Domino 208
│   │   │     │   │             Banner Domino 333
│   │   │     │   │                 Four Sq Banner 7M
│   │   │     │   │                     Justa Banner Guy 739
│   │   │     │   │                     │ Big Sky Guy
│   │   │     │   │                     │     LBCR Big Sky Spidele (1974)
│   │   │     │   │                     │ Justa Alpine 150Z
│   │   │     │   │ Chub Domino
│   │   │     │   │     Advd Chub Domino
│   │   │     │   │ Advd Seth Domino
│   │   │     │ Advanced Anxiety
│   │   │     │     Anxiety
│   │   │     │         Super Anxiety
│   │   │     │             SRR Dom Return
│   │   │     │ Advanced Domino 38
│   │   │     │   Victor Dom Return
│   │   │     │ Circle M Adv Domino
│   │   │     │   │ CMR Advance Domino
│   │   │     │   │     CMR Choice Domino
│   │   │     │   │         ALF Choice Domino 6 (1947)
│   │   │     │   │         │ SV Polled Lamplighter
│   │   │     │   │         │     SV Pld Lamplight 38
│   │   │     │   │         │         RF Domestic Lamp 19*
│   │   │     │   │         │             Morlunda Choice D67
│   │   │     │   │         │ SV Choice Domino 5
│   │   │     │   │         │ SV Choice Domino 107
│   │   │     │   │         │     SV Choice Mischief 9
│   │   │     │   │         │         OK Gold Mischief 20
│   │   │     │   │         │             EHF Advance Domino**
│   │   │     │   │         │ SV Beau Perfection
│   │   │     │   │ CMR Advance Dom 16
│   │   │     │   │     GB Advance Dom 26
│   │   │     │   │ CMR Advance Dom 19
│   │   │     │   │ CMR Adv Domino 35
│   │   │     │   │     Advance Domino 67
│   │   │     │   │ CMR Advance Dom 50
│   │   │     │   │ CMR Advance Dom 61
│   │   │     │   │     RCR Advance Larry
│   │   │     │   │ CMR Advance Dom 84
│   │   │     │   │     CMR Advance Dom 135
│   │   │     │   │ CMR Advance Dom 111
│   │   │     │       SRF Adv Domino 20
│   │   │     │           WEB&S Advance Dom 1
│   │   │     │               WEB&S Perfect Adv 2
│   │   │     │                   WEB&S Beau Domino 6
│   │   │     │                       PF Advance Blanchard
│   │   │     │                           MCF Adv Beau Perfect (1970)
│   │   │     Advance Domino 50
│   │   │     │ Prince Lad
│   │   │     │ Regulator Domino
```

(Continued on next page)

Polled Prince Domino Families (Continued)

```
Prince Domino
    Real Prince Domino
    Dandy Domino 2
    Choice Domino
    Prince Domino Mixer
    Prince Domino 1
    Prince Domino 4
    Dandy Domino
    Sarah Domino (1938)
    Prince Domino Misch
        Advance Domino
            Advance Domino 47
            Advance Fairview
                Polled Advance Dom
                    Mesa Domino
                        GHR Anxiety Dom 50
                            GHR Mesa Return 344
                                GHR Mesa Return 336
                                    JAO Mesa Grande
                                        OR Mesa 337 N53*
                                            Mis Finch Victra N42 (1973)
                        Mesa Royal 5
                            Beau Mesa Domino
                        GHR Mesa Domino 44
                            GHR Mesa Return 437
                            GHR Mesa Return 516
                        GHR Mesa Domino 76
                            OO Del Mesa 32
                    FLR Adv Domino 14
        Polled Anxiety 4
            Polled Prince Domino
                Polled Prince Dom 1
                    Golden Nugget
                        Gold Mine (1946)
                            OK Gold Mine 25
                            (Gold Colonel) 1,806
                            Gold Pride
                                Gold Pride 20
                                    A Gold Pride 7**
                                LR Gold Mischief
                                    HW Adv Gold Misch 2
                                    HW Adv Gold Misch 3
                            GS Gold Nugget 3
                                Gold Monarch 20
                                    ALF Monarch 37 (1956)
                            Gold Emblem
                            Gatesford Gold Mine
                            (Gold Crown) 741
                            OK Gold Captain
                            Gold Rocket
                                E Gold Rocket
                                    E Domestic Rocket 4
                            Gold Pilot
                            Gold Co-Pilot
                            Gold Choice
                            FF Gold Banner
                                FF Gold Prince
                                    FF Gold Prince Jr.
                                        (FF Gold Aster 94) 521
                            White Gold
                                HJW Gold Domino 16
                Anxiety Woodrow
```

(Continued on next page)

Polled Prince Domino Families (Continued)

Prince Domino
    | *Real Prince Domino*
    | *Dandy Domino 2*
    | *Choice Domino*
    | *Prince Domino Mixer*
    | *Prince Domino 1*
    | *Prince Domino 4*
    | *Dandy Domino*
    | *Sarah Domino (1938)*
    | *Prince Domino Misch*
        Advance Domino
            | *Advance Domino 47*
            | *Advance Fairview*
                Polled Advanced Dom
                    | *Mesa Domino*
                    | *FLR Adv Domino 14*
                Polled Anxiety 4
                    Polled Prince Domino
                    Anxiety Woodrow
                    | Polled Don Carlos
                        Polled Beau Mischief
                        Carlos Mischief
                        Carlos Mischief 4
                            | C Mischief Pres 8
                                C Mischief Pres 44
                                    P Mischief Domino
                            | G Adv Carlos Misch 2
                                DCF Carlos Misch 6
                                    DCF Larry Domino C (1952)
                                        ALF Carlos Rupert 4 (1955)
                    | Mellow Mischief
                        | T Mellow Misch Jr.
                        | T Mellow Conqueror 2
        Advance Mischief
            Advance Mischief 78
                Advance Mischief 2
                    Advance Mischief 64
                        Advance Mischief 3 (1946)
                            OK Seth Mischief
            Prince Adv Mischief
                Prince Mischief 55
                    Prince Mischief 11
                        Pine Park Mischief

(Continued on next page)

Polled Prince Domino Families (Continued)

```
Prince Domino
   │ Real Prince Domino
   │ Dandy Domino 2
   │ Choice Domino
   │ Prince Domino Mixer
   │ Prince Domino 1
   │ Prince Domino 4
   │ Dandy Domino
   │ Sarah Domino (1938)
   │ Prince Domino Misch
        │ Advance Domino
        │ Advance Mischief
             │ Advance Mischief 78
             │ Prince Adv Mischief
             │ Advance President
                  The Lamplighter
                     │ Domestic Lamp
                     │ Modest Lamplighter
                          │ Modest C Lamp 2
                               │ Beau Modest C
                                    MSF Modest C Lamp
                               │ HPM Lamplighter 3
                                  HPM Lamplighter 20
                                       │ W Kings Lamplighter
                                       │ BW Lamplighter 05
                                            W Lamplighter F26
                                       │ W Lamplighter D43
                                       │ W Lamplighter E43 (1965)
                                       │ BW Lamplighter 756
                                            SR Lamplighter 9**
                     Polled Mod Lamp 2
                          │ Advance Lamplighter
                             FLR Advance Lamp 1
                               FLR Advance Lamp 12
                                    │ PTZ Advance Anx 4
                                       Justa Apex Q479
                                            Beartooth Ponderosa (1971)
                                    │ WPHR Carlos Lamp (1960)
                                         │ WPHR Carlos Lamp Jr.
                                         │ WPHR Carlos Lamp 29
                          HHR Modest Mischief
                          Modest C Lamp 4
                               (Jr. Modest Lamp 4) 512
                                    │ (BB Beau Lamp 90) 610
                                    │ M Polled Anxiety
                                    │ Jr. Modest Lamp 6
                                         Mousels Lamplighter
                                              │ Oak Ridge Lamp 12***
                                                   │ Dia WH Lamp 22V
                                                        Mischief Lamp 13Y*
                                                   │ Dia WH Lamp 23W
                                                        Advancer 228D (1974)
                                              │ Oak Ridge Lamp 18
                                                   Kiyiwana New Trend*
                               Modest Jr. 11
                                    GHR Modest Mesa 1
                                         (GHR Mesa M178)* 757
                               Modest Jr. 17
                                    │ BS Lamplighter 1
                                    │ DIR Real Lamp 51
                                         Nulook Lamp 1A*
                               │ CPH Jr. Modest 3
                                    CPH Miss Modest 32 (1966)
                     Polled Mod Lamp 26
                     │ HHR Modest Lamp
```

(Continued on next page)

Polled Prince Domino Families (Continued)

```
Prince Domino
  │ Real Prince Domino
  │ Dandy Domino 2
  │ Choice Domino
  │ Prince Domino Mixer
  │ Prince Domino 1
  │ Prince Domino 4
  │ Dandy Domino
  │ Sarah Domino (1938)
  │ Prince Domino Misch
  │   │ Advance Domino
  │   │ Advance Mischief
  │       │ Advance Mischief 78
  │       │ Prince Adv Mischief
  │       │ Advance President
  │           The Lamplighter
  │             │ Domestic Lamp
  │                 │ Modest Lamplighter
  │                     │ Modest C Lamp 2
  │                         │ Beau Modest C
  │                         │ HPM Lamplighter 3
  │                     │ Polled Mod Lamp 2
  │                         │ Advanced Lamplighter
  │                         │ HHR Modest Mischief
  │                         │ Modest C Lamp 4
  │                         │ Polled Mod Lamp 26
  │                         │ HHR Modest Lamp
  │                         │ FLR Beau Modest
  │                             │ FLR Beau Modest 10
  │                                 │ E Beau Perfect
  │                                     │ Perfect Lamplighter
  │                                         │ HCJ Beau Perf Lamp 1*
  │                                 │ E Domestic Lamp
  │                                     │ E Apex Lamplighter
  │                         │ Choice Lamplighter
  │                             │ GJ Lamplighter 22
  │                             │ GJ Lamplighter 24
  │                     │ Polled Mod C Lamp 3
  │                     │ Polled Mod Lamp
  │                     │ Beau Lamplighter
  │                         │ FLR Beau Lamp 21
  │                             │ FLR Adv Modest Lamp
  │                                 (FLR Modest Lamp 4)* 552
  │                 │ Domestic Lamp 6
  │                     │ Atomic D Lamp
  │                         │ Atomic D Lamp 15
  │                             │ Polled Atomic Lamp
  │                                 │ MH Atomic Mischief
  │                 │ Domestic Lamp 35
  │                     │ Supreme Lamp 21
  │ Imperial Lamp
  │     │ Imperial Lamp 37
  │         │ Don Lamplighter
  │             │ Don Lamplighter 3
  │                 │ HPHR Lamplighter D8
```

BEAU MISCHIEF FAMILIES

```
Beau Mischief
    Anxiety Mischief
        Anxiety Jr.
            Super Anxiety
                Super Anxiety 5
                    Supreme Anxiety 7
                        Supreme Anxiety 10
                            Pierre Supreme 12
                                Pierre Supreme 40
        Anxiety 23
            Domestic Mischief
                Domestic Mischief 32
                    Merry Mischief 2 (1947)
                    Domestic Mischief 97
                        GR Mischief Blanco
                            C Domestic Misch 23
                                JFG Domes Misch 253
                                Diamond
                                    Golden Diamond (1961)
                                        Golden Reward
                                    Mister Diamond (1964)
                                    Helen Diamond R (1964)
                                    Diamond Too***
                                        Diamond Lil 416 (1965)
                                    Diamond Mine
                                        TF Diamond Mine Jr.
                                JFG Domes Misch 500
                                    LLF Miss Domes Lamp (1969)
                            JFG Domes Misch 26
                                JFG Domes Misch 202
                            JFG Domes Misch 32
                                M Domestic Misch 2
                                    VGHF Domes Misch 97
                Woodrow Mischief
                    Woodrow Misch 48
                        Cir T Wdrw Misch 8M
                            CV Wdrw Misc Lad 37U
                                CV Wdrow Anx Lad 5W
                                    (Craigview Misch 26Y) 832
                        Cir T Wdrw Misch 24J
                            Morlunda Fabulous
                Woodrow Mischief 3
                (Domestic Mischief 6) 577
                    Domestic Anxiety
                        Domestic Anxiety Jr.
                            Circle T Dom Anx 25F
                                Oak Ridge Dom Anx 93
                                    Justa Domes Guy 440Z
                                        QC 440 Sis 573C (1972)
                                    Justa Sterling 232Y
                                Oakland Ridge Anx 56*
                                    FLF Domestic Anx 6
                                    (Canadian Fantastic) 536
                                    Just Anx Murphy 286Y*
                                    Justa Anx Boldman 2T
                                        WSF Trumode Oakey*
                                Oakland Ridge Anx 73
                                Oakland Ridge Anx 78
                                    AAB Oakland Ridge*
```

(Continued on next page)

Beau Mischief Families (Continued)

```
Beau Mischief
   │ Anxiety Mischief
   │    │ Anxiety Jr.
   │    │ Anxiety 23
   │           Domestic Mischief
   │              │ Domestic Mischief 32
   │              │ Woodrow Mischief
   │              │ Woodrow Mischief 3
   │              │ (Domestic Mischief 6) 577
   │                    │ Domestic Anxiety
   │                    │    │ Domestic Anxiety Jr.
   │                    │    │ Domestic Anxiety 89
   │                    │          Domestic Anx 89 OG18
   │                    │             │ Domestic Anx 107E
   │                    │                  JDH Mischief Maker
   │                    │                     JDH Golden Anxiety**
   │                    │                        │ HH Victor Anxiety 35** (R)
   │                    │                        │ HH Victor Anxiety 42 (R)
   │                    │                        │ HH Victor Anxiety 62 (R)
   │                    │             │ Domestic Anxiety 29D
   │                    │                  │ Domes Anxiety 3610**
   │                    │                  │ OG Domes Anx 0344 (R)
   │                       Domestic Woodrow
   │                          │ Domestic Woodrow 23
   │                          │    HHR DW23 117
   │                          │ Domestic Woodrow 133
   │                          │ Essar Domestic W (1949)
   │                          │    WPHR Domestic W
   │                          │       (Domestic W14) 571
   │                          │          │ FLF Domes Maid 45 (1962)
   │                          │          │ Crail Domestic W7
   │                          │          │ GB Domestic Mixer 37
   │                          │ Domes Woodrow 120
   │                          │ HHR DW 70
   │                          │ HHR DW 85
   │                               (CMF Domestic Woodrow) 832
   │                                  │ (CPH Woodrow 41)* 777
   │                                  │ CPH Jr. Wood Lamp 2
   │                       Domestic Misch 259 (1950)
   │                          HHR DM259 14
   │                       Woodrow Mischief 2
   │                          Woodrow Mischief 21
   │                             Dahl Mischief 2
   │                    │ McH Domestic Nugget
   │    Beau Aster
   │       Jealous Aster
   │    Beau Blanchard
   │       Beau Blanchard 95
   │          Beau Blanchard 224
   │             Canadian Blanchard (1932)
   │    Mischief Mixer
   │       Mischief Mixer 28
   │          Mischief Mixer 5
   │             Colorado Misch J118
   │                CMR Dandy Mixer
   │                   CMR Dandy Mixer 12
   │ Battle Mischief
   │    Battle Mischief Jr.
   │       Battle Mischief 7
   │          │ Battle Domino 5
   │          │ Battle Domino 11
   │          │ Rose Battle 22 (1942)
   │          │ Rose Battle 47 (1944)
   │          │ ALF Rose Battle 78 (1945)
```

VARIATION FAMILIES

```
Variation
    | Excellent Ion
    |     | Formation 2
    |     |     | Bullion 4
    |     |     |     | Don Bullion 2
    |     |     |     |     Don Bullion Jr. (1929)
    |     |     |     | Bullion Garfield (1923, 1925)
    |     |     |     | Bullion 7
    |     |     |     |     | Model Bullion Jr.
    |     |     |     |     |     Model Dale
    |     |     |     |     |         Model Russell
    |     |     |     |     |             Russell Woodford
    |     |     |     |     |                 | Bullion Repeater
    |     |     |     |     |                 |     KD Captain Vagabond
    |     |     |     |     |                 | Russell Dale 10
    |     |     |     |     |                       KD Dale Woodford
    |     |     |     | Model Bullion (1927)
    |     | Pawnee Chief
    |     |     | Pawnee Rollo
    |     |     |     | Pawnee Rollo 9
    |     |     |     |     | Rollos Boy 8
    |     |     |     |     |     | Prince Rollo 1 (1934)
    |     |     |     |     |     |     | Pawnee Druid
    |     |     |     |     |     |     |     Pawnee Druid 8
    |     |     |     |     |     |     |         Pawnee Druid 20
    |     |     |     |     |     |     |             Pawnee Druid 30
    |     |     |     |     |     |     |                 BPF Pawnee Druid
    |     |     |     |     |     |     |                     BPF Pawnee Mixer
    |     |     |     |     |     |     |                         | Pawnee Perfect
    |     |     |     |     |     |     |                         | GF Pawnee Mixer 14*
    |     |     |     |     |     |     |                         |     | FLF Winston Mixer 3
    |     |     |     |     |     |     |                         |     | FLF Winston Mix 710
    |     |     |     |     |     |     |                         |     | PS Winston Mixer 32
    |     |     |     |     |     |     |                         |         PS Triple Mixer 101
    |     |     |     |     |     |     |                         |             SSS Triple Mixer 42 (R)
    |     |     |     |     |     |     |                         | (BPF Pawnee Mixer 24) 534
    |     |     |     |     |     |     |                         |     | PS Pawnee Mixer 133
    |     |     |     |     |     |     |                         |     | PS Modest Mixer 65
    |     |     |     |     |     |     |                         |     | PS Pawnee Mixer 162 (1963)
    |     |     |     |     |     |     |                         |     | MVF Lady Pawnee M6 (1963)
    |     |     |     |     |     |     |                         | BPF Pawnee Mixer 21
    |     |     |     |     |     |     |                               PS Domestic Mixer
    |     |     |     |     |     | Olga Domino (1940)
    |     |     |     |     | Pawnee Rollo 10 (1933)
    |     |     |     |     | Pocahontas (1934)
    |     | Gemmation 2
    |     |     | Gemmation 32
    |     |     |     | Miss Larks Gem
    |     |     |     |     | Canadian Gem (1930)
    |     |     |     |     |     Belle Canadian (1933)
    |     |     |     |     | Miss Donald Gem (1926)
    |     |     | Valicias Gem (1924-25)
    | Variation 3
    |     Foundation 4
    |         Foundation 25 (1924)
```

POLLED PRESIDENT FAMILIES

```
Polled President
     President Mischief
          President Mischief 9
          President Misch 22
               HHR Mischief Duke
                    CEK Zato Mischief (1957)
                    HHR Mischief Duke 01
                         HHR Misch Duke 01A
                         CEK Royal Domino
                              CEK Royal Lady 12 (1959)
                         CEK Zato Tone
                              CEK Zato Tonette 1 (1958)
     Polled President Jr.
          T Dom President 1
               G Advance President
                    SS Adv President
                         S Domino Pres 31
                    G Jr. Adv President
                    G Jr. Adv President 2
          T Plato President 40
               Rollos Pawnee
                    WW Royal Rollo 13
                         WW Fairview 3
                    WW Royal Rollo 40
```

POLLED ZATO FAMILY

```
Beau Brummel
     Beau Donald
          Prince Rupert
               Rupert Donald
                    Sir Rupert
                         Hazford Rupert
                              Hazford Rupert 25
                                   Zato Rupert
                                        Zato Tone 2
                                             H&D Zato Tone Lad 8
                                                  H&D Tone Lad 105
                                                       TR Zato Heir
                                                            TR Zato Heir 88
                                                                 (TR Royal Zato 27) 565
                                                                      (Polled Zato Heir 27) 581
                                                                      HDR Polled Zato 6
                                                                           FF Golden Zato 3
                                                                      HDR Polled Zato 21
                                                                      HDR Polled Zato 28
                                                                      HDR Polled Zato 63
                                                                      HDR Polled Zato 67
                                                                           G&R Perfect Mixer***
                                                            TR Zato Heir 181
                                                                 PS Polled Zato
```

BOCALDO FAMILY

```
Anxiety 4
     Don Carlos
          Beau Brummel
               Painter
                    Caldo 2
                         Bocaldo
                              Bocaldo 6
                                   Beautys Bocaldo
                                   | Marys Bocaldo
                                   | Jimmy Bocaldo 2
                                        Bocaldos Real 2
                                             Polled Bo Real
                                                  PPHR Bocaldo Real 2
                                                       PPHR Bocaldo Real 38
                                                            PPHR Trailblazer 3
                                                                 PPHR Trailblazer 12
```

BEAU PERFECTION/WORTHMORE FAMILIES

```
Polled Plato
| Polled Dad 21
|     Polled Perfection
|          Beau Perfection
|               Beau Perfection 2
|                    Beau Perfection 100
|                         Beau Perfect 231
|                              Perfection Lad 5
|                         Beau Perfection 234
|                              Beau Perfect 246
|                                   ALF Beau Rollo
|                                        DM Rollo Domino 1
|                                        | DT Rollo Domino
|                                        | WW Choice Rollo
|                                   (ALF Beau Rollo 11) 500
|                                        | HCJ Beau Rollo 51
|                                        |     HCJ Beaurollo Lmp 12*
|                                        | Beau Rollo Lamp 2
|                                   ALF Beau Rollo 56
|                                        Coastal B Rollo 34
|                                             Coastal B Rollo 100
|                                                  (Coastal B Rollo 165) 545
|                                   ALF Beau Rollo 77
|                                        HH Beau Rollo 12
|                                   ALF Stella Beau 7 (1946)
|                                   Perfect Domino 1
|                                        | Perfect Aster
|                                        | GPH Perfect Aster 82
|                                             (GPH Perf Aster 155)* 878
|                                   Perfect Domino 3
|                                        ALF Perfect Domino 3
|                                             RF Perfect Domino 40
|                                                  RF Perf Domino 4003
|                                                       RF Perf Domino A19
|                                                            WW Perfect Domino
|                                   4E Beau Rollo 11
|                                        FHF Beau Rollo 33
|                                             W Beau Rollo 10*
|                         Beau Perfect 263
|                              Beau Battle
|                                   Veo Battle Mixer
|                                        Pawnee Beau Perfect
|                                        | FLF Perfect Mixer 30*
|                                        |     CPH Beau Mischief 3 (1969)
|                                        | BPF Pawnee Perfect
|
```

(Continued on next page)

Beau Perfection/Worthmore Families (Continued)

```
Polled Plato
  | Polled Dad
  |     Polled Perfection
  |         Beau Perfection
  |             Beau Perfection 2
  |                 Beau Perfection 100
  |                     Beau Perfection 234
  |                       | Beau Perfect 231
  |                       | Beau Perfect 246
  |                       | Beau Perfect 263
  |                       | Beau Perfect 281
  |                             EER Beau Perfect
  |                                 EER Beau Perfect 15
  |                                     LD Beau Mesa Prince
  |                                         MPHR Beau Mode 2
  |                                           | Tizhome Beau Mode 1U
  |                                           |   | TF Beau Mode 13A*
  |                                           |   | TF Beau Mode 23X*
  |                                           |       GK Klondike Gainer (R)
  |                                           |       GK Klondike Gainer 3**(R)
  |                                           |     | Revs Beau Lad 25D (1973)
  |                                         Tizhome Beau Mode 6S
  |                                             (Canam Investor) 762 (1968)
  |                                               | RR Miss Canam OD39 (1970)
  |                                               | Miss Investor (1971)
  |                                               | GK Investors Leader**
  |                                         | Tizhome Beaumode 16S
  |                                             (CR Beau Mode 26Y) 678
  | Polled Star
        Polled Success
            Worthmore
                Worthmore 1
                    Joe
                        Worthmores Beau
                            Worthmores Beau 3
                                Worthmores Beau Jr.
                                    Worthmores Beau Jr. 2 (1938-39)
                                      | Worthmores Return 2
                                      | PVF Beau Advance
                                            Advancemore Mischief
                                      | PVF Advance Worth 2 (1942-43)
```

BONNY B DOMINO/COLONEL DOMINO FAMILIES

```
Giant
   Polled Admiral 2
   |  Polled Ito
   |     Polled William
   |        Bonnie Real
   |           Real Mischief
   |              Curly Mischief
   |                 Mischief Domino
   |                    Spidel B54
   |                       Spidel 110K
   |                          Spidel 753O
   |                             Spidel 400C
   |                                Spidel 600H
   |                                   Spidel 175
   |                                      Roundup Domino 3L
   |                                         (Predominant 25U)* 813
   |                                            | (Justamere R Ian 176X)* 575
   |                                            PRL Roundup Cal C183*
   |                                            PRL B Dame 153Y (1968)
   |        Bonnie Russell (1922)
   |        Bonnie Russell 2
   |           Marvels Bonnie (1928)
   |           |  Tommy Stanway
   |           |     Husky Domino
   |           |        Don Domino (1936)
   |           Marvels Bonnie Jr.
   |              Beau Bonnie 56
   |                 Colonel Domino
   |                    | HSF Beau Domino 2
   |                    Hazford Domino 3
   |                    BCF Colonel Royal
   |     Polled Peach 10
   |        Polled Dale
   |           Russell Dale
   |              | Princess Dale (1928)
   |              Princess Dale 2 (1929)
   |  Echo Grove
   |     Glendale Echo
   |        Woodrow Grove
   |           Kendale 24
   |              Bonny Blanchard 54
   |                 Bonny B1
   |                    | Bonny B 2 Jr.
   |                    |    Voigts Bonny B28
   |                    Bonny B Domino
   |                       Bonny B Domino 1
   |                          Bonny B Mischief 11
   |     Belle Grove 5 (1922)
   |     Lady Beatrice (1923)
   |     Echo Mischief
   |        Prime Mischief (1926)
   |     Polled Quality
   |        Polled Plato
   |           Polled Plato 8
   Royal Giant
      Progressor
         Joe Smith
            Joe Smith 19
               Woodrow
                  Woodrow 12
                     Woodrow 29
   Prince of Iowa
      Jolly
         Wizard Alex
            Wizard Alex 15
               Model Alex (1931)
```

Pedigree of Giant
The Fountain Head Sire of Polled Herefords

GIANT (1 Polled)
101740
May 3, 1899
O. F. Nelson

McKinley (horned) 74548

- Sir Frank 61465
  - Stone Mason 42393
    - Stone Mason 9 29071 — Beau Real 11055 / Stonebrook Lady 19488
    - Grace Imperial 34246 — Prince Imperial 2 6054 / Grace 2 19820
  - Victoria 43490
    - Famous 15170 — Edmund 6553 / Queen of the Prairie 1400
    - Bonita 8159 — Governor 4 1293 / Bonny Lass 8 8160
- Vashti 59208
  - Arkell 46598
    - Governor Simpson 19646 — Anxiety 4 9904 / Lynnette 13547
    - Lily 11441 — Basil Duke 11437 / Myrtle 5 10468
  - Virtue 49479
    - Plump 37400 — Lord Portland 11195 / Rosalind 15291
    - Viola 2921 — Ranger 1222 / Verbena 1225

Jess (horned) 35808

- Winfield 29710
  - Victor Emanuel 23165
    - Merry Monarch 5794 — Triumph 3 5795 / Peach 3600
    - Sawgate 14465 — Othello 8246 / Crosscut 14466
  - Coquette 6332
    - Waxwork 6320 — The Grove 3 2490 / Waxy 6321
    - Ceres 6317 — Joe Sterling 1197 / Cornelia 10350
- Alice Adams 4709
  - Prince of Adams 1463
    - Sir Arthur 705 — Sir Charles 543 / Hebe 2 694
    - Tulip 1423 — Northern Star 592 / Victoria 3 826
  - Lula 1460
    - Gold Dust 1461 — Bristol Bill 555 / Edith 574
    - Fredona First 1459 — Fair Boy 2 570 / Laura 1458

Pedigree of Variation
Second Sire of Note in Polled Hereford History

VARIATION
(14 Polled) 152699
Aug. 26, 1902
John G. Thomas

Milwaukee 79198

- Royal 62065
  - Le Roy 45892
    - Western Eagle 28109 — Anxiety 4 9904 / Gertrude 17261
    - Lizzie 36266 — Peeping Tom 3 15409 / Laurel 2 17272
  - Pansy 52502
    - Wamba 47566 — Heriot 1 24729 / Maud 2 6908
    - Dorothy 34621 — Dictator 2 6505 / Dorcas Wilton 21429
- Peerless Gipsy 63642
  - Peerless Wilton 12774
    - Garfield 7015 — Quickset 6853 / Plum 7016
    - Peerless 10902 — Lord Wilton 4057 / Delight 6959
  - Gipsy Girl 39217
    - Star Grove 10 17431 — The Grove 3 2490 / Annie Clark 1306
    - Queen of the Gipsies 20508 — Dauphin 18 3368 / Queen of the Lilies 7263

Wilton Velvet 79201

- Peerless Wilton 12774
  - Garfield 7015
    - Quickset 6853 — Regulus 3849 / Spot 3339
    - Plum 7016 — Challenge 1561 / Plum 7017
  - Peerless 10902
    - Lord Wilton 4057 — Sir Roger 3850 / Lady Claire 4116
    - Delight 6959 — Sir Frank 2674 / Eva 6960
- Jazel 60540
  - Sanhedrim 46180
    - Star Grove 19 26594 — The Grove 3 2490 / Annie Clark 1306
    - The Grove Maid 18 26571 — The Grove 3 2490 / Lovely 5375
  - Jingle 45619
    - Peerless Wilton 12774 — Garfield 7015 / Peerless 10902
    - Jessie Clark 2 10916 — Anxiety 3 4466 / Jessie Clark 7368

National Champions and Top Sellers
1916-1974

1916

Polled Harmon 455755-4571, by Polled Ito. Top selling bull at $1,430—James Paul, Friendship, Ind., to W. H. Campbell, Grand River, Iowa—57 head averaged $476.

Polled Marvel 445481-3142, by Polled Pride. Top selling female at $1,005—Glaves & Painter, Lewistown, Mo., to Painter Bros., Stronghurst, Ill.

1917

Echo Mischief 552693-7789, by Echo Grove. Top selling bull at $5,000—Glendale Stock Farm, Aspen, Colo., to John J. Kelleher, Patterson, Iowa—41 head averaged $846.

Polled Marvel 2 535373-7148, by Polled Pride. Top selling female at $1,525—E. H. Gifford, Lewiston, Neb., to Glaves & Painter, Lewistown, Mo.

1918

Polled Repeater 2 602680-10646, by Repeater 8. Top selling bull at $4,000—Glendale Stock Farm, Aspen, Colo., to G. M. Ross & Son, Ross, Iowa—42 head averaged $1,230.

Musoda Gem 566887-8860, by Gemmation 2. Top selling female at $2,500—G. E. Pettigrew & Son, Flandreau, S. D., to H. N. Vaughn, Stronghurst, Ill.

1919

Fairmount 637588-13936, by Bullion 4. Top selling bull at $2,025—W. O. Modlin, Upland, Ind., to Forbes & Neal, Marshall, Minn.—50 head averaged $1,122.

Pearl 676756-11873, by Polled Plato. Top selling female at $2,625—Grube & Scherzer, Larned, Kan., to Henry and H. J. Smith, Octavia, Neb.

1920

Fair Foundation 752485-19843 (changed to Beau Admiral 894496-19843), by Polled William. Top selling bull at $4,550—Glaves & Painter, Lewistown, Mo., to C. E. Brown, Rushville, Ill.—90 head averaged $962.

Princess Mischief 733542-19371, by Echo Mischief. Top selling female at $3,000—T.

F. Kelleher Estate, Des Moines, Iowa, to Goernandt Bros., Aurora, Kan.

1921

Curtis Repeater 826156-19297, by Polled Repeater. Top selling bull at $1,100—H. J. Smith, Octavia, Neb., to D. D. Pargin, Pridra, Colo.—181 head averaged $346.

Pearl 676756-11873, by Polled Plato. Top selling female at $3,500—J. D. Brunton, Aspen, Colo., to Renner, Person & Wilkey, Terre Haute, Ind.

1922

Bonnie Russell 957566-26306, by Polled William. Champion bull shown by R. C. Glaves, Lewistown, Mo.

Belle Grove 5 931866-25848, by Echo Grove. Champion female shown by Glendale Stock Farm, Aspen, Colo.

Marvels Anxiety 2 1014580-31906, by Marvels Pride 2. Top selling bull at $600—Ralph Painter, Stronghurst, Ill., to E. H. Vajnar, Chelsea, Iowa—101 head averaged $221.

Polled Daisette R 850152-21138, by Wizard Alex. Top selling female at $615—George T. Rew, Silver City, Iowa, to George D. Keith, Wichita Falls, Texas.

1923

Bullion Garfield 632729-14806, by Bullion 4. Champion bull shown by Renner Stock Farm, Hartford City, Ind.

Lady Beatrice 876967-23823, by Echo Grove. Champion female shown by Glendale Stock Farm, Aspen, Colo.

Marvels Dandy 1087166-37491, by Marvels Pride 2. Top selling bull at $2,000—Ralph Painter, Stronghurst, Ill., to Mt. Larcon Pastoral Co., Euroa, Gladstone, Qld., Australia—130 head averaged $219.

Princes Beauty 707345-18081, by Hartford Prince. Top selling female at $440—J. W. Chumley, New Tazewell, Tenn., to R. C. Glaves, Lewistown, Mo.

1924

Foundation 25 1000291-34844, by Foundation 4. Champion bull shown by Renner Stock Farm, Hartford City, Ind.

Valicas Gem 952956-27558, by Gemmation 2. Champion female shown by G. E. Pettigrew & Son, Flandreau, S. D.

Bonnie Real 1120597-42094, by Polled William. Top selling bull at $1,100—R. C. Glaves, Lewistown, Mo., to Mt. Larcon Pastoral Co., Euroa, Gladstone, Qld., Australia—56 head averaged $215.

Oh Lula 742938-18257, by Polled Quality. Top selling female at $235—Ray H. Swope, Clarion, Iowa, to Webster County Farm, Ft. Dodge, Iowa.

1925

Bullion Garfield 632729-14806, by Bullion 4. Champion bull shown by Renner Stock Farm, Hartford City, Ind.

Valicas Gem 952956-27558, by Gemmation 2. Champion female shown by G. E. Pettigrew & Son, Flandreau, S. D.

Russells Pride 1304003-45732, by Bonnie Russell. Top selling bull at $600—O. S. Wilson, Canton, Mo., to C. H. Zybell, Lake City, Iowa—69 head averaged $170.

Princess Gem 1222105-42498, by Gemmation 2. Top selling female at $185—G. E. Pettigrew & Son, Flandreau, S. D., to W. R. Elliott, Taylor Ridge, Ill.

1926

Prime Mischief 1112395-36734, by Echo Mischief. Champion bull shown by John J. Kelleher, Patterson, Iowa.

Miss Donald Gem 1225091-42431, by Miss Larks Gem. Champion female shown by P. M. Christenson & Son, Lone Rock, Iowa.

Prime Mischief 1112395-36734, by Echo Mischief. Top selling bull at $615—John J. Kelleher, Patterson, Iowa, to Gilligan Bros., Graff, Iowa—60 head averaged $245.

Miss Donald Gem 1225091-42431, by Miss Larks Gem. Top selling female at $345—P. M. Christenson & Son, Lone Rock, Iowa, to Clausen Bros., Schaller, Iowa.

1927

Model Bullion 1374897-48042, by Bullion 7. Champion bull shown by W. A. Wilkey & Co., Sullivan, Ind.

Perfection Gem 2 1245389-43374, by Gemmation 2. Champion female shown by G. E. Pettigrew & Son, Flandreau, S. D.

Priam 1492772-54016, by Bonnie Real. Top selling bull at $855—John J. Kelleher, Patterson, Iowa, to William Heydeman, Hartley, Southern Rhodesia, Africa—80 head averaged $212.

Perfection Gem 2 1245389-43374, by Gemmation 2. Top selling female at $410—G. E. Pettigrew & Son, Flandreau, S. D., to William Heydeman, Hartley, Southern Rhodesia, Africa.

1928

Marvels Bonnie 1499884-54122, by Bonnie Russell 2. Champion bull shown by R. C. Glaves, Lewistown, Mo.

Princess Dale 1393251-50536, by Russell Dale. Champion female shown by F. R. Mullendore & Son, Franklin, Ind.

Marvels Bonnie 1499884-54122, by Bonnie Russell 2. Top selling bull at $1,225—R. C. Glaves, Lewistown, Mo., to Burleson & Johns, Whitney, Texas—45 head averaged $320.

Fern Grove 1492774-54018, by Bonnie Real. Top selling female at $400—John J. Kelleher, Patterson, Iowa, to Roberts Loan & Cattle Co., Roundup, Mont.

1929

Don Bullion Jr. 1340323-47693, by Don Bullion 2. Champion bull shown by Charles Riffe & Sons, Windfall, Ind.

Princess Dale 2 1474998-54239, by Russell Dale. Champion female shown by F. R. Mullendore & Son, Franklin, Ind.

Bullion 93 1635503-60651, by Bullion Repeater. Top selling bull at $800—Robert Galbraith, What Cheer, Iowa, to William Spidel, Roundup, Mont.—66 head averaged $257.

Miss Willdale 7 1550827-58377, by Willdales Best. Top selling female at $500—O. S. Wilson, Canton, Mo., to Johnson Bros., Jacksboro, Texas.

1930

Canadian Gem 1647647-61730, by Miss Larks Gem. Champion bull shown by P. M. Christenson & Son, Lone Rock, Iowa.

Miss Fairfax 1623062-60800, by Fair Bullion. Champion female shown by J. L. Curran, Mason City, Iowa.

Beau Blanchard 46 1740906-65709, by Beau Blanchard. Top selling bull at $600—J. E. and George C. Kirstein, Clarion, Iowa, to

P. Cambier, Orange City, Iowa—49 head averaged $307.

Princess Dale 2 1474998-54239, by Russell Dale. Top selling female at $515—F. R. Mullendore & Son, Franklin, Ind., to Johnson Bros., Jacksboro, Texas.

1931

Model Alex 1648401-60947, by Wizard Alex 15. Champion bull shown by W. S. Wescott, Woodbine, Iowa.

Iowa Violet 1755258-64676, by Violets Boy. Champion female shown by J. L. Curran, Mason City, Iowa.

Marvels Corrector 1850848-71740, by Curley Marvel. Top selling bull at $550—F. E. Painter, Stronghurst, Ill., to Usher Bros., Scollard, Alta.—45 head averaged $226.

Jean Mischief 1789851-69862, by Bright Mischief. Top selling female at $300—W. H. Campbell & Son, Grand River, Iowa, to Hodgson Bros., Ottawa, Ill.

1932

Canadian Blanchard 1728630-65502, by Beau Blanchard 224. Champion bull shown by P. M. Christenson & Son, Lone Rock, Iowa.

Iowa Bell 1807908-67118, by Violets Boy. Champion female shown by J. L. Curran, Mason City, Iowa.

Duchess Pride 1979931-78476, by Painters Pride. Top selling bull at $327.50—Ralph Painter, Stronghurst, Ill., to R. C. Burgoin, Silver City, Iowa—38 head averaged $96.

1933

Pawnee Rollo 10 1884854-72303, by Pawnee Rollo 9. Champion bull shown by Hugh H. White, Keller, Texas.

Belle Canadian 2033586-78819, by Canadian Gem. Champion female shown by P. M. Christenson & Son, Lone Rock, Iowa.

Pawnee Rollo 11 1941104-75500, by Pawnee Rollo 9. Top selling bull at $200—Hugh H. White, Keller, Texas, to T. B. Bailey, Austin, Texas—32 head averaged $89.

Miss Henrietta 1684759-83051, by Polled Plato 56. Top selling female at $110—Burleson & Johns, Whitney, Texas, to L. D. Wythe, Weatherford, Texas.

1934

Prince Rollo 1 2076623-81921, by Rollos Boy 8. Champion bull shown by Hugh H. White, Keller, Texas.

Pocahontas 2104282-83195, by Pawnee Rollo 9. Champion female shown by Hugh H. White, Keller, Texas.

Choice Domino Plus 2 2082369-86149, by Choice Domino 18. Top selling bull at $610—Paul Bize, Julian, Neb., to L. A. Schreiner, Kerrville, Texas—25 head averaged $266.

Cecile Plato 2034072-79961, by Mossy Plato 3. Top selling female at $1,500—George Trenfield, Follett, Texas, to Hackney Bros., McGregor, Texas.

1935
(No National Show Held in 1935)

Prime Bullion 2655432-92184, by Donald Bullion. Top selling bull at $315—Hodgson Bros., Ottawa, Ill., to George Ross & Son, Gray, Iowa—39 head averaged $153.

Babe Bullion 1808536-69716, by Bocaldo A. Top selling female at $185—W. E. Wardal, Northwood, Iowa, to Ernest Combs, Good Hope, Ill.

1936
(Two National Sales Held in 1936)

Don Domino 2449779-102022, by Husky Domino. Champion bull shown by R. C. Glaves, Lewistown, Mo.

Curley Ann 2267009-95601, by Canadian Gem 3. Champion female shown by P. M. Christenson & Son, Lone Rock, Iowa.

Santos Peter 2159251-87657, by Santos Bullion. Top selling bull at $725—J. L. Curran, Mason City, Iowa, to Carlos Fournier, Montevideo, Uruguay—55 head averaged $159.

Lady Domino 2329262-96932, by Victor Domino. Top selling female at $200—N. M. Leonard, Waukee, Iowa, to Brownell Combs, Lexington, Ky.

Prince Domino Gem 2 2471107-104307, by Prince Domino Gem. Top selling bull at $600—Louis J. Marzen, Marble Rock, Iowa, to Andrew R. Joughin, Arroyo Grande, Calif.—60 head averaged $177.

Beaus Ruth 2459790-99809, by Beau Blanchard 2. Top selling female at $225—G. A. Treiber, Hebron, N. D., to Worner Polled Hereford Farms, San Jose, Ill.

1937

Domino Gem 2454812-103497, by Mac Domino. Champion bull shown by P. M. Christenson & Son, Lone Rock, Iowa.

Lucille Canadian 2564652-106048, by Dale Canadian. Champion female shown by P. M. Christenson & Son, Lone Rock, Iowa.

Santos Jim 2476611-106133, by Santos Peter. Top selling bull at $525—J. L. Curran, Mason City, Iowa, to James Sparkes, Lyndley, Australia—49 head averaged $156.

Happy Bullion 8 2462359-103816, by Anxiety Bullion 11. Top selling female at $170—Worner Polled Hereford Farms, San Jose, Ill., to Brownell Combs, Lexington, Ky.

1938

Worthmores Beau Jr. 2 2421065-98791, by Worthmores Beau Jr. Champion bull shown by Jesse Riffel & Sons, Enterprise, Kan.

Sarah Domino 2367351-104787, by Prince Domino. Champion female shown by P. M. Christenson & Son, Lone Rock, Iowa.

Domino Blanchard 2 2711221-117627, by Prince Domino 94. Top selling bull at $650—George Ross & Son, Gray, Iowa, to Earl Blanchard, Oshkosh, Neb.—58 head averaged $179.

Dora Domino 2705689-117622, by Prince Domino 94. Top selling female at $300—George Ross & Son, Gray, Iowa, to J. F. Crutcher, Henning, Tenn.

1939

Worthmores Beau Jr. 2 2421065-98791, by Worthmores Beau Jr. Champion bull shown by Jesse Riffel & Sons, Enterprise, Kan.

Rosella 2872580-127702, by Advanced Domino 30. Champion female shown by Orvil E. Kuhlmann, North Platte, Neb.

Double Advanced 2920394-127792 (changed to Myrtlewood Domino), by Advanced Domino 30. Top selling bull at $730—Orvil E. Kuhlmann, North Platte, Neb., to Brownell Combs, Lexington, Ky.—77 head averaged $218.

Rosella 2872580-127702, by Advanced Domino 30. Top selling female at $600—Orvil E. Kuhlmann, North Platte, Neb., to M. P. Moore, Senatobia, Miss.

1940

Real Plato Domino 2839351-123565, by Plato Domino 1. Champion bull shown by Leslie Brannan, Timken, Kan.

Olga Domino 2751764-118784, by Prince Rollo 1. Champion female shown by M. P. Moore, Senatobia, Miss.

Aster Domino 3105406-137952, by Advanced Domino 30. Top selling bull at $2,000—Orvil E. Kuhlmann, North Platte, Neb., to E. L. Cord, Dyer, Nev.—122 head averaged $328.

Martha Mischief 2886399-128305, by Domestic Mischief. Top selling female at $625—Halbert & Hoggett, Mertzon, Texas, to E. L. Cord, Dyer, Nev.

1941

CMR Rollo Domino 3254000-147094, by Victor Domino 4. Champion bull shown by M. P. Moore, Senatobia, Miss.

Stella 3027156-131313, by Advanced Domino 30. Champion female shown by Orvil E. Kuhlmann, North Platte, Neb.

Anxiety Aster 3146090-149495, by Mossy R Domino 1. Top selling bull at $2,600—John E. Rice, Lodge Grass, Mont., to Carlos Fournier, Montevideo, Uruguay—166 head averaged $480.

Rose Battle 12 3252519-152716, by Battle Domino 5. Top selling female at $1,150—John M. Lewis & Sons, Larned, Kan., to E. L. Cord, Dyer, Nev.

1942

PVF Advance Worth 2 3370036-161144, by Worthmores Beau Jr. 2. Champion bull shown by Jesse Riffel & Sons, Enterprise, Kan.

Rose Battle 22 3452561-165588, by Battle Domino 5. Champion female shown by John M. Lewis & Sons, Larned, Kan.

Home Maker 37 3430558-163918, by Home Maker 2. Top selling bull at $2,700—John E. Rice, Lodge Grass, Mont., to Fowler Farms, Chattanooga, Tenn.—167 head averaged $573.

Rose Battle 22 3452561-165588, by Battle Domino 5. Top selling female at $1,500—John M. Lewis & Sons, Larned, Kan., to M. P. Moore, Senatobia, Miss.

1943

PVF Advance Worth 2 3370036-161144, by Worthmores Beau Jr. 2. Champion bull shown by Jesse Riffel & Sons, Enterprise, Kan.

D Victoria Domino 8 3332243-163646, by Victor Domino 126. Champion female shown by M. P. Moore, Senatobia, Miss.

Real Plato Domino 25 3685934-182746, by Real Plato Domino. Top selling bull at $3,500—John M. Lewis & Sons, Larned, Kan., to Allgood & McDaniel, Liberty, S. C.—168 head averaged $769.

Rose Battle 36 3502582-170325, by Battle Domino 5. Top selling female at $3,550—John M. Lewis & Sons, Larned, Kan., to Seco Farms, Arcadia, Mo.

1944

Trumode Domino 8 3759927-185723, by Plato Domino 36. Champion bull shown by John E. Rice, Lodge Grass, Mont.

Rose Battle 47 3717533-186844, by Battle Domino 5. Champion female shown by John M. Lewis & Sons, Larned, Kan.

Bonnie B Domino 2 3989587-204134, by Bonnie B Domino. Top selling bull at $5,025—Joe Weedon, Grosvenor, Texas, to P. P. Santayana, Montevideo, Uruguay—176 head averaged $777.

Autumn A8 4070886-203305, by Battle Domino 11. Top selling female at $6,000—Welch Bros., Garfield, Kan., to M. P. Moore, Senatobia, Miss.

1945

CMR Rollo Domino 28 4087774-207222, by CMR Rollo Domino. Champion bull shown by M. P. Moore, Senatobia, Miss.

ALF Rose Battle 78 4076741-216431, by Battle Domino 5. Champion female shown by John M. Lewis & Sons, Larned, Kan.

Mischief Lamplighter 2 4104187-208469, by Woodrow Mischief 2. Top selling bull at $7,000—R. A. Halbert, Sonora, Texas, to Henry and Jean Holfeldt, Belair, Md.—137 head averaged $1,014.

ALF Stella Beau 7 4067076-217806, by Beau Perfect 246. Top selling female at $6,100—John M. Lewis & Sons, Larned, Kan., to M. P. Moore, Senatobia, Miss.

1946

(Three Nationals Held in 1946)

Advance Mischief 3 4317801-231483, by Advance Mischief 64. Champion bull (Baton Rouge, La., and Des Moines, Iowa) shown by R. A. Halbert, Sonora, Texas.

ALF Stella Beau 7 4067076-217806, by Beau Perfect 246. Champion female (Baton Rouge, La., Des Moines, Iowa, and Columbus, Ohio) shown by M. P. Moore, Senatobia, Miss.

Gold Mine 4567141-251156, by Golden Nugget. Champion bull (Columbus, Ohio) shown by Orvil E. Kuhlmann, North Platte, Neb.

Royal Beau Domino 4310395-233310, by Beau Domino 283. Top selling bull (Baton Rouge, La.) at $5,000—Hervale Farms, Wayne, Neb., to J. R. Reeves, Clarkesville, Ga.—60 head averaged $756.

ALF Stella Beau 17 4314322-228633, by Beau Perfect 246. Top selling female (Baton Rouge, La.) at $2,900—John M. Lewis & Sons, Larned, Kan., to R. A. Halbert, Sonora, Texas.

Star Domino 2 4351816-231522, by Star Domino 72. Top selling bull (Columbus, Ohio) at $3,475—M. P. Moore, Senatobia, Miss., to G. Bernard Fenwick, Glyndon, Md.—62 head averaged $671.

Polly Mischief 4235802-222433, by Domestic Mischief. Top selling female (Columbus, Ohio) at $2,950—Jim Gill, Whon, Texas, to Gentry D. Adams & Son, Allendale, Ill.

ALF Real Domino 12 4616252-256861, by Battle Domino 37. Co-top selling bull (Des Moines, Iowa) at $1,850—John M. Lewis & Sons, Larned, Kan., to Harry Manfull, Gettysburg, S. D.—63 head averaged $577.

C Domino President 96 4424068-245312, by T Domino President 5. Co-top selling bull (Des Moines, Iowa) at $1,850—P. C. Campbell, Temple, Okla., to Hajek Bros., Odell, Neb.

T Alene Comprest 2 4633521-258862, by Advance Comprest. Top selling female (Des Moines, Iowa) at $2,125—George Trenfield & Son, Follett, Texas, to Peckerwood Ranch, Mt. Vernon, Ill.

1947

ALF Choice Domino 6 4695708-267299, by CMR Choice Domino. Champion bull shown by John M. Lewis & Sons, Larned, Kan.

Merry Mischief 2 4650080-258605, by Domestic Mischief 32. Champion female shown by Jim and Fay Gill, Coleman, Texas.

ALF Choice Domino 6 4695708-267299, by CMR Choice Domino. Top selling bull at $35,000—John M. Lewis & Sons, Larned, Kan., to A. G. Rolfe, Poolesville, Md.—97 head averaged $1,550.

ALF Bluebonnet 17 4695737-269108, by CMR Choice Domino. Top selling female at $4,200—John M. Lewis & Sons, Larned, Kan., to R. M. Ivens, Johnson City, Tenn.

1948

Numode 29 4924655-279184, by Trumode Domino 8. Champion bull shown by John E. Rice, Sheridan, Wyo.

CMR Blanche Domino 25 5334733-316117, by CMR Rollo Domino 12. Champion female shown by M. P. Moore, Senatobia, Miss.

ALF Choice Domino 35 5206629-306759, by CMR Choice Domino. Top selling bull at $9,200—John M. Lewis & Sons, Larned, Kan., to Paul Greening, Pomona, Calif.—133 head averaged $1,262.

ALF Lady Mix 1 4933873-287507, by ALF Pawnee Mixer 21. Top selling female at $5,450—John M. Lewis & Sons, Larned, Kan., to Brownell Combs, Lexington, Ky.

1949

Essar Domestic W 5402208-324611, by Domestic Woodrow. Champion bull shown by P. C. Campbell, Temple, Okla.

CMR Blanche Domino 25 5334733-316117, by CMR Rollo Domino 12. Champion female shown by M. P. Moore, Senatobia, Miss.

ALF Perfect Domino 3 5561791-331784, by Perfect Domino 3. Top selling bull at $5,750—John M. Lewis & Sons, Larned, Kan., to Riverdale Farms, Charlottesville, Va.—74 head averaged $1,252.

ALF Lady Mix 6 5382556-321368, by ALF Pawnee Mixer 21. Top selling female at $6,000—John M. Lewis & Sons, Larned, Kan., to C. C. Potter, Pottstown, Pa.

1950

Domestic Mischief 259 5750466-360949, by Domestic Mischief 6. Champion bull shown by R. A. Halbert, Sonora, Texas.

7-Up Royal Maid 13 6004256-396422, by Numode 16. Champion female shown by William I. Moore, Banner, Wyo.

Domestic Woodrow 244 5750462-360942,

by Domestic Woodrow 4. Top selling bull at $7,000—R. A. Halbert, Sonora, Texas, to Hayfields Farm, Cockeysville, Md.—72 head averaged $1,755.

ALF Stella Beau 66 5791065-362266, by Beau Perfect 246. Top selling female at $8,100—John M. Lewis & Sons, Larned, Kan., to F. E. Crosslin & Son, Eagleville, Tenn.

1951

SF Larry Mischief 7 5843832-368704, by SFL Domino Mischief. Champion bull shown by Sumter Farm & Stock Co., Geiger, Ala.

EER Victoria Tone 22 6106299-390310, by EER Victor Domino 12. Champion female shown by E. E. Moore, Senatobia, Miss.

SF Larry Mischief 7 5843832-368704, by SFL Domino Mischief. Top selling bull at $20,000—Sumter Farm & Stock Co., Geiger, Ala., to O'Bryan Ranch, Hiattville, Kan.—75 head averaged $2,576.

HHR Miss DW71 6266281-414511, by Domestic Woodrow. Top selling female at $4,100—Halbert & Fawcett, Sonora, Texas, to Henry and Jean Holfeldt, Bel Air, Md.

1952

DCF Larry Domino C 6588698-500000, by DCF Carlos Misch 6. Champion bull shown by John M. Lewis & Sons, Larned, Kan.

Baca Queen 503520, by RHR Baca Duke. Champion female shown by William Caton, Corydon, Ky.

7-Up Royal Mode 46, by Numode 16. Top selling bull at $7,000—William I. Moore, Banner, Wyo., to Rock Hill Ranch, Walls, Miss.—102 head averaged $1,259.

Baca Queen 503520. Top selling female at $7,500—William Caton, Corydon, Ky., to Blue Crystal Farms, Hillards, Ohio.

1953

ALF Battle Mixer 30 7024524-600000, by ALF Pawnee Mixer 24. Champion bull shown by John M. Lewis & Sons, Larned, Kan.

EER Victoria Tone 50 517382, by EER Victor Domino 12. Champion female shown by E. E. Moore, Senatobia, Miss.

FLR Advance Lamp 3, by Advance Lamplighter. Top selling bull at $7,650—F. L. Robinson & Son, Kearney, Neb., to P. H. Ginsbach, Dell Rapids, S. D.—58 head averaged $1,469.

ALF Lady Return 69, by ALF Beau Mixer 3. Top selling female at $4,600—John M. Lewis & Sons, Larned, Kan., to C. M. Cubbage, Port Monmouth, N. J.

1954

JFG Domestic Mixer P7584118-700000, by BR Proud Mixer. Champion bull shown by Jim and Fay Gill, Coleman, Texas, and John M. Lewis & Sons, Larned, Kan.

JR Dandymaid B12 678199, by Double Dandymode 5. Champion female shown by John E. Rice & Sons, Sheridan, Wyo.

HHR Mischief Duke 01A, by HHR Mischief Duke 01. Top selling bull at $9,000—Halbert & Fawcett, Miller, Mo., to M&O Polled Herefords, Worthington, Ind.—92 head averaged $895.

JR Dandymaid 13, by Double Dandymode. Top selling female at $3,150—John E. Rice & Sons, Sheridan, Wyo., to Wagon Wheel Ranch, Germantown, Tenn.

1955

ALF Carlos Rupert 4 P8219331-800000, by DCF Larry Domino C. Champion bull shown by John M. Lewis & Sons, Larned, Kan.

EER Victor Duchess 18 713820, by EER Victor Duke. Champion female shown by E. E. Moore, Senatobia, Miss.

CEK Larry Domino 133, by MW Larry Domino 133. Top selling bull at $8,200—Mr. & Mrs. C. E. Knowlton, Bellefontaine, Ohio, to A. G. Rolfe, Poolesville, Md.—49 head averaged $1,100.

ALF Lady Monarch 17, by Gold Monarch 20. Top selling female at $2,500—John M. Lewis & Sons, Larned, Kan., to Holly Springs Farm, Covington, Ga.

1956

ALF Monarch 37 P8898842-900000, by Gold Monarch 20. Champion bull shown by Hanson Hereford Farm, Red Wing, Minn.

Bay Lady Tontine 14 867824, by ALF Battle Mixer 30. Champion female shown by Otis H. Smith, Lewes, Del.

ALF Monarch 35, by Gold Monarch 20. Top selling bull at $15,200—John M. Lewis & Sons, Larned, Kan., to Circle A Ranch, Sandersville, Ga.—59 head averaged $1,728.

ALF Lady Battle 3 741262, by ALF Battle Mixer 30. Top selling female at $3,750—John M. Lewis & Sons, Larned, Kan., to W. H. Lewis, Greenwood, Ark.

1957

CEK Zato Mischief P9682112-1000000, by HHR Mischief Duke. Champion bull shown by Mr. & Mrs. C. E. Knowlton, Bellefontaine, Ohio.

Miss Larry Carlos 20, by WW Larry Carlos. Champion female shown by Falklands Farm, Schellsburg, Pa.

CEK Zato Mischief P9682112-1000000, by HHR Mischief Duke. Top selling bull at $10,000—Mr. & Mrs. C. E. Knowlton, Bellefontaine, Ohio, to A. G. Rolfe, Poolesville, Md.—49 head averaged $1,325.

Gatesford N Lady 48, by Gatesford Numode. Top selling female at $3,400—Pleasant Point Plantation, Beaufort, S. C., to A. D. Davis, Alachua, Fla.

1958

CEK Mixer Return P10022952-1100000, by ALF Mixer Return 115. Champion bull shown by O'Bryan Ranch, Hiattville, Kan.; Mr. & Mrs. C. E. Knowlton, Bellefontaine, Ohio; and Huber Ranch, Schneider, Ind.

CEK Zato Tonette 1 982947, by CEK Zato Tone. Champion female shown by Mr. & Mrs. C. E. Knowlton, Bellefontaine, Ohio.

CEK Pawnee Mixer, by ALF Mixer Return 115. Top selling bull at $56,000—Mr. & Mrs. C. E. Knowlton, Bellefontaine, Ohio, to Tine W. Davis, Montgomery, Ala.—49 head averaged $2,379.

CEK Zato Tonette 1 982947, by CEK Zato Tone. Top selling female at $3,700—Mr. & Mrs. C. E. Knowlton, Bellefontaine, Ohio, to Tom and Ruth Durbin, Mt. Pleasant, Ohio.

1959

WCH Misch R Royal 1 P10026177-1200000, by CMR Mischief Domino 101. Champion bull shown by Huber Ranch, Schneider, Ind.

CEK Royal Lady 12 1075135, by CEK Royal Domino. Champion female shown by Mr. & Mrs. C. E. Knowlton, Bellefontaine, Ohio.

WCH Misch R Royal 1 P10026177-1200000, by CMR Mischief Domino 101. Top selling bull at $7,000—Huber Ranch, Schneider, Ind., to Coleson Place, Magnolia, Ark.—48 head averaged $1,350.

E1 Modest Bulova P10140734-1064339, by E Gold Rocket. Top selling female at $3,500—W. J. Largent, Folsom, N. M. and Glendon H. Etling, Gruver, Texas, to John Wahl, Enumclaw, Wash.

1960
(No National Sale Held in 1960)

WPHR Carlos Lamp P10760052-1300000, by FLR Advance Lamp 12. Champion bull shown by Walton Polled Hereford Ranch, Akron, Colo., and Pitts Polled Herefords, Dixon, Mont.

W Numaid Henrietta 2, by JR Numode J43. Champion female shown by Hunsinger Hereford Farm, Mt. Eaton, Ohio.

1961
Golden Diamond P11389703-1400000, by Diamond. Champion bull shown by Orvil E. Kuhlmann & Sons, North Platte, Neb.

CEK Lady Return 69, by CEK Mixer Return. Champion female shown by Mr. & Mrs. C. E. Knowlton, Bellefontaine, Ohio.

GF Pawnee Mixer 14 P11138333-1241355, by BPF Pawnee Mixer. Top selling bull at $7,500—Norman Greenway, LaGrangeville, N. Y., TO Millcreek Valley Farm, Johnstown, Pa.—59 head averaged $1,225.

CLR Miss Roltrend 77 P11574686-1383278, by CMR Rollo Mixer. Top selling female at $2,505—W. H. Lewis, Greenwood, Ark., to John N. Brown, Louisville, Ky.

1962
HAB Victor Mischief P11357684-1343009, by Gay Hills Victor L42. Champion bull shown by H. A. Bartholomew, Washington, D. C., and James C. Linthicum, Dayton, Md.

FLF Domestic Maid 45 1501237, by Domestic W14. Champion female shown by Falklands Farm, Schellsburg, Pa.

HAB Victor Mischief P11357684-1343009, by Gay Hills Victor L42. Top selling bull at $102,000—H. A. Bartholomew, Washington, D. C., and James C. Linthicum, Dayton, Md., to Rogers Farms, Inc., Morton, Miss., and Otis H. Smith, Lewes, Del.—44 head averaged $3,908.

CSR Miss Woodrow 3 P11895506-1455299, by CMF Domestic Woodrow. Reserve sale

champion and top selling female at $5,000—Carnation Farms, Carnation, Wash., and Clark Properties, Inc., Carnation, Wash., to Rogers Farms, Inc., Morton, Miss.

1963
PS Pawnee Mixer 162 P11754481-1455370, by BPF Pawnee Mixer 24. Champion bull shown by Pennsylvania State University, University Park, Pa.; Millcreek Valley Farm, Johnstown, Pa.; and Michigan State University, East Lansing, Mich.

MVF Lady Pawnee M6 P11919779-1444268, by BPF Pawnee Mixer 24. Champion female shown by Scott Hereford Farm, Hickory Flat, Miss.

HR Brae Mixer X12732078, by CEK Mixer Return. Sale champion and top selling bull at $7,700 (one-half interest)—Howard Riechmann, Valmeyer, Ill., to Cumberland Trail Farms, St. Elmo, Ill.—59½ head averaged $1,525.

McF Miss Superol 3 P12056134-1529716, by CMR Superol 64. Top selling female at $3,500—McCullough Farm, Hardinsburg, Ind., to Scott Hereford Farm, Hickory Flat, Miss.

1964
Mister Diamond X12547644, by Diamond. Champion bull shown by Orvil E. Kuhlmann & Sons, North Platte, Neb.

Helen Diamond R X12547652, by Diamond. Champion female shown by Orvil E. Kuhlmann & Sons, North Platte, Neb.

BW Junior Lamp B103 P12392353-1548878, by HPM Lamplighter 20. Sale champion and top selling bull at $10,900—Wolfe Hereford Ranch, Wallowa, Ore., to Emory Moore Ranch, Kings Valley, Ore.—48 lots averaged $1,337.

FF Merrietta 30 X12607292, by FF Golden Zato 3. Sale champion and top selling female at $2,550—Foley Farms, Ltd., Middletown, Calif., to Glen Corning, Bellingham, Wash.

1965
W Lamplighter E43 X13078274, by HPM Lamplighter 20. Champion bull shown by Wolfe Hereford Ranch, Wallowa, Ore.

Diamond Lil 416 X13431825, by Diamond Too. Champion female shown by Hervale Farms, Wayne, Neb.

OK Diamond 85 X12846346, by Diamond. Sale champion and top selling bull at $12,500—Orvil E. Kuhlmann & Sons, North Platte, Neb., to Kelowna Ranches, Ltd., Kelowna, B. C.—69 head averaged $1,563.

Diamond Lil 416 X13431825, by Diamond Too. Top selling female at $5,000— Hervale Farms, Wayne, Neb., to Kelowna Ranches, Ltd., Kelowna, B. C.

1966

AAB Superol X13742631, by SFR-CEK Superol. Champion bull shown by Anthony A. Buford, Caledonia, Mo.

CPH Miss Modest 32 X13411896, by CPH Jr. Modest 3. Champion female shown by Pennsylvania State University, University Park, Pa.

AAB Superol X13742631, by SFR-CEK Superol. Sale champion and top selling bull at $19,000 (one-half interest)—Anthony A. Buford, Caledonia, Mo., to Mont-Vue Farm, Niota, Tenn.—45½ head averaged $2,853.

PS Winnie Mixer 96 X13864954, by GF Pawnee Mixer 14. Top selling female at $3,700—Pennsylvania State University, University Park, Pa., to Rogers Farms, Inc., Morton, Miss.

1967

ALF Beau Perfect X13767544, by D Mixer Domino 36. Champion bull shown by John M. Lewis & Sons, Larned, Kan.

AAB Miss Roltrend 82 X14086818, by CLR Rollotrend A. Champion female shown by Anthony A. Buford, Caledonia, Mo.

PS Triple Mixer 101 X13864945, by PS Winston Mixer 32. Top selling bull at $28,000 (one-half interest)—Pennsylvania State University, University Park, Pa., and Sam Sells & Son, Moultrie, Ga., to Baughman Ranches, Inc., Paradise, Kan.—64 head averaged $2,754.

PS Winnie Mixer 106 X14167494, by GF Pawnee Mixer 14. Top selling female at $4,400—Pennsylvania State University, University Park, Pa., to Brookhill Farm, Clarksville, Mo.

1969 (Houston)

Canam Investor X17998550, by Tizhome Beau Mode 6S. Grand champion bull shown by Glenkirk Farms, Maysville, Mo.

PRL B Dame 153Y X20084050, by Predominant 25U. Grand champion female shown by Bedford Farms, Inc., Shelbyville, Tenn.

Canam Investor X17998550, by Tizhome Beau Mode 6S. Top selling bull at $40,000 (one-fourth interest)—Glenkirk Farms, Maysville, Mo., to Rollins Ranches, Atlanta, Ga.—63¼ head averaged $2,552.

PS Modern Miss 149 X14660157, by PS Modest Mixer 65. Reserve sale champion and top selling female at $5,000—Pennsylvania State University, University Park, Pa., to Queenie Creek Ranch, Vermillion, Alta.

1969 (Atlanta)

CPH Beau Mischief 3 X14755669, by FLF Perfect Mixer 30. Grand champion bull shown by Falklands Farm, Schellsburg, Pa.

LLF Miss Domes Lamp X14791185, by JFG Domes Misch 500. Grand champion female shown by Lucky Lane Farm, Hillsboro, Ohio.

CPH Beau Mischief 3 X14755669, by FLF Perfect Mixer 30. Top selling bull at $17,000 (one-fourth interest)—Connolly Polled Herefords, St. Helena, Calif.; Storm Ranch, Dripping Springs, Texas; and Falklands Farm, Schellsburg, Pa., to Windsweep Farm, Thomaston, Ga.—47½ head averaged $2,352.

ALF Lady Lamp 33 X2000035, by HCJ Beau Perf Lamp 1. Sale champion and top selling female at $2,500—John M. Lewis & Sons, Larned, Kan., to Carroll A. Smith, Lookeba, Okla.

1970

(no National held this year)

1971

MCF Adv Beau Perfect by PF Advance Blanchard, Grand Champion Bull, shown by McCaskill Farms, Timewell, Ill.

RR Miss Canam OD39 by Canam Investor, Grand Champion Female, shown by Glenkirk Farms, Maysville, Mo., and Rollins Ranch, Atlanta, Ga.

Justamere Alpine 150Z, top selling bull consigned by Justamere Farms, Ltd., Lloydminster, Sask.—$16,000 for one-fourth interest—32¼ head averaged $1,945.

WHF Miss Mesa Dom 08, co-top selling

female consigned by Jerry Ballard & Sons, Nashport, Ohio—$2,500.

SQ D Anuity V Lass 16, co-top selling female consigned by Murray Duke, Langbank, Sask.—$2,500.

1972

Beartooth Ponderosa by Justa Apex Q479, Grand Champion Bull, shown by Beartooth Ranch, Columbus, Mont.

Miss Investor by Canam Investor, Grand Champion Female, shown by Jesse Riffel & Sons, Enterprise, Kan.

RHR Mesa Lamp 07, top selling bull consigned by Reed Ranch, Savonburg, Kan., and TL Custom Fitting Service, Osage City, Kan.—$10,400—21 head averaged $3,706.

GK Investress Queen, top selling female consigned by Glenkirk Farms, Maysville, Mo.—$5,000.

1973

Victorious K47 U81 by RWJ Victor F18 K47, Grand Champion Bull, shown by Falklands Farm, Schellsburg, Pa.

QC 440 Sis 573C by Justa Domes Guy 440Z, Grand Champion Female, shown by Beartooth Ranch, Columbus, Mont.

Victorious K47 U81, top selling bull consigned by Justamere Farms, Ltd., Lloydminster, Sask.; Jack A. Oleson, Avon & Longmont, Colo.; and Falklands Farm, Schellsburg, Pa.—$31,000 for one-fourth interest—19¾ head averaged $6,813.

QC 440 Sis 573C, top selling female consigned by Beartooth Ranch, Columbus, Mont.—$6,600.

1974 (Denver)

Revs Beau Lad 25D by TF Beau Mode 23X, Grand Champion Bull, shown by Ivan G. Fulford, Pelly, Sask., Johnson Bros., Spy Hill, Sask., and Little Beaver Creek Ranches, Missoula, Mont.

Mis Finch Victra N42 by OR Mesa 337 N53, Grand Champion Female, shown by Glen & Lillian Allen, Hillsboro, Tex., and John S. Finch, Corsicana, Tex.

Playmaker 4, top selling bull consigned by T. J. Griswold, Livingston, Wis.—$38,500 for one-fourth interest—16 head averaged $12,470.

SFR Vicki Domette 6, top selling female consigned by Beartooth Ranch, Columbus, Mont., and Santa Fe River Ranch, Alachua, Fla.—$5,000.

1974 (Louisville)

Advancer 228D by Dia WH Lamplight 23W, Grand Champion Bull, shown by Beartooth Ranch, Columbus, Mont., Ellis Farms, Chrisman, Ill., Glenview Farms, Inc., Farmington, Ill., and Wooden Shoe Farms, Friday Harbor, Wash.

LBCR Big Sky Spidele by Big Sky Guy, Grand Champion Female, shown by Little Beaver Creek Ranches, Missoula, Mont.

Klondike 23X 249D, top selling bull consigned by Fernwood Farm, Barrington, Ill.; Glenkirk Farms, Maysville, Mo.; Glenview Farms, Inc., Farmington, Ill.; and Klondike Farms, Ltd., Douglas, Man.—$25,000 for one-fourth interest—20½ head averaged $6,002.

LBCR Big Sky Spidele, top selling female consigned by Little Beaver Creek Ranches, Missoula, Mont.—$20,000.

National Champions and Top Sellers

The following are photographs that were available from the American Polled Hereford Association files on most of the national champion and top selling bulls and females listed on the preceding pages. From the first National sale in 1916 and the first show in 1922, these cattle were judged the finest of their day.

Polled Harmon—1916 top selling bull

Polled Marvel—1916 top selling female

Echo Mischief—1917 top selling bull

Fairmount—1919 top selling bull

Fair Foundation—1920 top selling bull

Princess Mischief—1920 top selling female

Pearl—1919 and 1921 top selling female

Bonnie Russell—1922 champion bull

Belle Grove 5—1922 champion female

Polled Daisette R—1922 top selling female

Lady Beatrice—1923 champion female

Marvels Dandy—1923 top selling bull

Foundation 25—1924 champion bull

Valicas Gem—1924 and 1925 champion female

Bonnie Real—1924 top selling bull

Bullion Garfield—1923 and 1925 champion bull

Russells Pride—1925 top selling bull

Prime Mischief—1926 champion and top selling bull

Miss Donald Gem—1926 champion and top selling female

Model Bullion—1927 champion bull

Perfection Gem 2—1927 champion and top selling female

Priam—1927 top selling bull

Marvels Bonnie—1928 champion and top selling
bull

Princess Dale—1928 champion female

Don Bullion Jr.—1929 champion bull

Princess Dale 2—1929 champion and 1930 top
selling female

Bullion 93—1929 top selling bull

Miss Willdale 7—1929 top selling female

Canadian Gem—1930 champion bull

Miss Fairfax—1930 champion female

Beau Blanchard 46—1930 top selling bull

Model Alex—1931 champion bull

Iowa Violet—1931 champion female

Marvels Corrector—1931 top selling bull

Jean Mischief—1931 top selling female

Iowa Bell—1932 champion female

Pawnee Rollo 10—1933 champion bull

Belle Canadian—1933 champion female

Pawnee Rollo 11—1933 top selling bull

Prince Rollo 1—1934 champion bull

Pocahontas—1934 champion female

Choice Domino Plus 2—1934 top selling bull

Cecile Plato—1934 top selling female

Don Domino—1936 champion bull

Curley Ann—1936 champion female

Prince Domino Gem 2—1936 top selling bull

Domino Gem—1937 champion bull

Lucille Canadian—1937 champion female

Santos Jim—1937 top selling bull

Sarah Domino—1938 champion female

Worthmores Beau Jr. 2—1938 and 1939 champion bull

Rosella—1939 champion and top selling female

Real Plato Domino—1940 champion bull

Olga Domino—1940 champion female

CMR Rollo Domino—1941 champion bull

Stella—1941 champion female

PVF Advance Worth 2—1942 and 1943 champion bull

Rose Battle 22—1942 champion and top selling female

D Victoria Domino 8—1943 champion female

Real Plato Domino 25—1943 top selling bull

Trumode Domino 8—1944 champion bull

Rose Battle 47—1944 champion female

CMR Rollo Domino 28—1945 champion bull

ALF Rose Battle 78—1945 champion female

Advance Mischief 3—1946 champion bull

ALF Stella Beau 7—1945 top selling and 1946 champion female

Gold Mine—1946 champion bull

ALF Choice Domino 6—1947 champion and top selling bull

Merry Mischief 2—1947 champion female

Numode 29—1948 champion bull

CMR Blanche Domino 25—1948 and 1949 champion female

ALF Choice Domino 35—1948 top selling bull

Essar Domestic W—1949 champion bull

ALF Perfect Domino 3—1949 top selling bull

Domestic Mischief 259—1950 champion bull

7-Up Royal Maid 13—1950 champion female

Domestic Woodrow 244—1950 top selling bull

ALF Stella Beau 66—1950 top selling female

SF Larry Mischief 7—1951 champion and top selling bull

EER Victoria Tone 22—1951 champion female

DCF Larry Domino C—1952 champion bull

Baca Queen—1952 champion and top selling female

ALF Battle Mixer 30—1953 champion bull

EER Victoria Tone 50—1953 champion female

JFG Domestic Mixer—1954 champion bull

HHR Mischief Duke 01A—1954 top selling bull

ALF Carlos Rupert 4—1955 champion bull

EER Victor Duchess 18—1955 champion female

ALF Monarch 37—1956 champion bull

Bay Lady Tontine 14—1956 champion female

ALF Monarch 35—1956 top selling bull

CEK Zato Mischief—1957 champion and top selling bull

Miss Larry Carlos 20—1957 champion female

CEK Mixer Return—1958 champion bull

CEK Zato Tonette 1—1958 champion and top selling female

CEK Pawnee Mixer—1958 top selling bull

WCH Misch R Royal 1—1959 champion and top selling bull

CEK Royal Lady 12—1959 champion female

WPHR Carlos Lamp—1960 champion bull

W Numaid Henrietta 2—1960 champion female

Golden Diamond—1961 champion bull

CEK Lady Return 69—1961 champion female

GF Pawnee Mixer 14—1961 top selling bull

HAB Victor Mischief—1962 champion and top selling bull

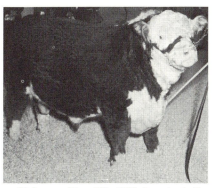

PS Pawnee Mixer 162—1963 champion bull

MVF Lady Pawnee M6—1963 champion female

Helen Diamond R—1964 champion female

BW Junior Lamp B103—1964 top selling bull

W Lamplighter E43—1965 champion bull

Diamond Lil 416—1965 champion and top selling female

AAB Superol—1966 champion and top selling bull

ALF Beau Perfect—1967 champion bull

AAB Miss Roltrend 82—1967 champion female

PS Triple Mixer 101—1967 top selling bull

Canam Investor—1969 (Houston) champion and top selling bull

PRL B Dame 153Y—1969 (Houston) champion female

PS Modern Miss 149—1969 (Houston) top selling female

LLF Miss Domes Lamp—1969 (Atlanta) champion female

CPH Beau Mischief 3—1969 (Atlanta) champion and top selling bull

MCF Adv Beau Perfect—1971 champion bull

RR Miss Canam OD39—1971 champion female

Beartooth Ponderosa—1972 champion bull

Miss Investor—1972 champion female

Victorious K47 U81—1973 champion and top selling bull

QC 440 Sis 573C—1973 champion and top selling female

Revs Beau Lad 25D—1974 (Denver) champion bull

Mis Finch Victra N42—1974 (Denver) champion female

Advancer 228D—1974 (Louisville) champion bull

LBCR Big Sky Spidele—1974 (Louisville) champion and top selling female